THE LAST BUTTERFLIES

THE LAST BUTTERFLIES

A SCIENTIST'S QUEST TO SAVE A RARE AND VANISHING CREATURE

NICK HADDAD

PRINCETON UNIVERSITY PRESS
PRINCETON AND OXFORD

Copyright © 2019 by Princeton University Press

Published by Princeton University Press
41 William Street, Princeton, New Jersey 08540
6 Oxford Street, Woodstock, Oxfordshire OX20 1TR

press.princeton.edu

All Rights Reserved

Library of Congress Control Number: 2018965563
ISBN: 978-0-691-16500-4

British Library Cataloging-in-Publication Data is available

Editorial: Alison Kalett and Kristin Zodrow
Production Editorial: Ellen Foos
Jacket Design: Chris Ferrante
Production: Erin Suydam
Publicity: Sara Henning-Stout and Julia Hall
Copyeditor: Amy K. Hughes

Jacket image: Bay Checkerspot by Kim Davis and Mike Stangeland

This book has been composed in Adobe Text Pro and Futura PT

Printed in the United States of America

10 9 8 7 6 5 4 3 2 1

TO KATHRYN, HELEN, AND OWEN

CONTENTS

	Preface	IX
1	A Sliver of Creation	1

PART I. THE RAREST BUTTERFLIES

2	Bay Checkerspot	21
3	Fender's Blue	41
4	Crystal Skipper	65
5	Miami Blue	85
6	St. Francis' Satyr	113
7	Schaus' Swallowtail	144

PART II. THE FLIGHT PATH FORWARD

8	The Final Flight of the British Large Blue	173
9	Monarchs: The Perils for Abundant Butterflies	187
10	The Last Butterfly?	202

Acknowledgments	217
Notes	221
Illustration Credits	239
Index	241

PREFACE

The unique butterfly with which I have been associated has perhaps the best latinized name in all of science. *Inglorius mediocris*—loosely translated as the Mediocre Skipper—is small, with wings that measure about one inch across (Plate 1, bottom). It is brown, a color broken by a few specks the size and hue of sand grains. One could argue that this butterfly, among the thousands of plants and animals that might be termed *mediocre* in appearance, deserves its name. I would like to think that the name is comical (and perhaps it is), but more importantly, to me, the Mediocre Skipper was my first real contribution to the study of rare butterflies.

I never set out to be a scientist or a steward of the world's rarest butterflies. I was not a young butterfly enthusiast. I never had a butterfly collection, and I do not remember raising caterpillars. I did not find my passion until later in life. Following my graduation from Stanford University in 1992, I decided to take a job with the Center for Conservation Biology at Stanford to inventory butterflies in northern Guatemala. At the time, I wasn't particularly drawn to butterflies, but work in Guatemala sounded exciting. Several of the center's butterfly biologists accompanied me for the first week. Then they left me in a tropical forest with a tent, a vintage mountain bike, and a butterfly net (Plate 1, top). They did not trust me, a young student, to identify the area's five-hundred-plus tropical butterfly species. So they left me with one instruction: collect all butterflies and send them to experts. I did so for two years.

Shortly after completing that task, I left Guatemala to begin graduate school. Five years passed, and George Austin, a museum scientist and a member of the group that took me to Guatemala, sent me one of his recently published papers. In it, he described some of the novel discoveries contained within the boxes of butterflies I had sent him.

One stood out. I learned that we had collected a butterfly that was a species new to science, *Inglorius mediocris*. At first that did not seem so novel. After all, we'd collected other new species, including *Calephelis tikal*, which was named after Tikal National Park, where we caught it during our first week of netting. But as I read the article I learned that the Mediocre Skipper was a more unusual find than that butterfly from Tikal. *Inglorius* was a new genus, a lineage of butterflies potentially millions of years old. (A genus is a grouping of similar species that are related by morphology and genetics; for example, the Monarch is in the genus *Danaus*, and there are twelve other species in that genus.) I was astonished that the collection I'd mailed George contained something unique. In the two decades since, others have collected only five more individuals of this species in Guatemala and bordering countries. No other species of the *Inglorius* genus have yet been discovered. The Mediocre Skipper impressed on me how much remains to be discovered about the diversity and the potential rarity of butterflies and of insects more generally.

I did not study rare butterflies before or in the few years after my research in Guatemala. My honors thesis as an undergraduate student was actually on birds found in oak forest, as I sought to understand how forest loss affected the types of birds that remained. My research sites, in California's San Francisco Bay Area, were immediately adjacent to the grasslands inhabited by a rare butterfly, the Bay Checkerspot (*Euphydryas editha bayensis*); however, I never saw it there.

I also did not study rare butterflies during graduate school at the University of Georgia. I entered graduate school to continue research on birds. With them in mind, I studied an approach that is used in biodiversity conservation to help the species of an area overcome the negative effects of habitat loss. Landscape corridors are "highways" for plants and animals that reconnect habitats that have been lost and fragmented. I worked with the US Forest Service to create a large experiment to test the corridors' effects. Shortly after I started this effort, my graduate adviser, Professor Ron Pulliam, nudged me toward studies of common butterfly species. Over the next decade, I found that landscape corridors were superhighways for butterflies and increased butterfly abundances. Yet none of the butterflies I studied was rare.

I peg the start of my search for rare butterflies to 2001. I was a new professor at North Carolina State University. At that time, government agencies and conservation organizations were struggling to develop conservation plans for the state's two rare butterflies, the St. Francis' Satyr (*Neonympha mitchellii francisci*) and the Crystal Skipper (*Atrytonopsis quinteri*). They asked me to bring my expertise on habitat loss and on butterflies to the conservation effort. I willingly embraced this opportunity. I found these butterflies to be especially enticing because they lived in heavily fragmented environments. Apparently, landscape corridors would speed their recovery. These butterflies and their environments were just what I'd been looking for, a conservation setting in which I could put into practice what I'd learned from my PhD research. As a naive over-optimist, I thought that my scientific expertise was just the ingredient needed to promote rapid recovery. I was going to be these species' savior.

A few years after I moved to North Carolina and started my studies of rare butterflies, I had a debate with one of my graduate students, Allison Leidner, about who was studying the rarest butterfly. Was it I, studying the St. Francis' Satyr, restricted to thirty-five acres on an artillery range in one army installation? Or was it Allison, who was studying the Crystal Skipper, restricted to a thirty-mile-long by 150-foot-wide strip of sand dunes on a barrier island? I am competitive, and I was determined to win this debate. The army thwarted my efforts, as at that time they would not let me into the artillery range to count the St. Francis' Satyrs. As I could not win, I moved on to find a different butterfly to study that might have been even rarer. My decades-long search for the rarest butterfly had begun.

I have studied rare butterflies for nearly two decades. It turns out that the special window opened by the rarest butterflies has given my science and worldview a unique perspective. These rare butterflies reveal the diversity of life and the potential for loss. Butterflies are the best-known group of insects, and their loss portends that of many other insects that comprise the vast majority of earth's diversity (excluding microorganisms). In this book, I stitch together stories about the discovery, science, threats, and conservation of the rarest butterflies in the world. While writing these chapters, I sought general insights into causes of butterfly decline and prospects for these insects' conservation and recovery. The lessons I extracted are applicable to other rare plants and animals, not only to butterflies.

THE LAST BUTTERFLIES

CHAPTER 1

A SLIVER OF CREATION

After I started my job at North Carolina State University in summer 2001, I toured the habitats of two of the last known populations of the St. Francis' Satyr (*Neonympha mitchellii francisci*), located at Fort Bragg army installation in southern North Carolina. I traveled through pinewoods on rutted dirt roads to visit a just-discovered population where the butterfly was easy to spot. This marked the beginning of my research on the St. Francis' Satyr. I had some early successes and found a few new populations. This inspired me to continue my search for undiscovered populations in remote wetlands at Fort Bragg. Every summer since, I have trudged through swamps and broken through walls of shrubs and vines. For the most part, my effort has been in vain. My challenges in finding new populations of St. Francis' Satyr were emblematic of the science and the search for the rarest butterfly that lay before me.

The butterflies that are the subject of this book represent just a sliver of creation. If we were at a dinner party and I asked you to think of a rare animal, what would come to mind? I would expect to hear names of animals such as the Giant Panda (*Ailuropoda melanoleuca*), the Black Rhinoceros (*Diceros bicornis*), or the Northern Spotted Owl (*Strix occidentalis caurina*). Like the butterflies I study, these animals are rare and threatened. Unlike the rarest butterflies, they are large, charismatic vertebrates. These animals are also different in a way that may not be immediately apparent (at least it was not to me): they are not nearly as rare as the rarest butterfly.

Rare butterflies make up a small number of earth's nineteen thousand or so butterfly species, and butterflies in general make up a small fraction of the estimated 5.5 million insect species. Relative to other insects, butterflies hold an advantage: they provide us clearer avenues for understanding general threats to biodiversity and pathways to conservation. We know much more about butterflies—their diversity, ecology, and evolution—than any other group of insect. We also know more about the size of their populations and about the area of their ranges—which means there are data to support my assessment of rarity.

Imagine the increasingly likely scenario in which you could corral all the living adults of all the very rarest butterflies and then hold them in your hands. If, for example, you could hold the entire world population of adult Schaus' Swallowtail (*Heraclides aristodemus ponceanus*) butterflies, its weight would be roughly six ounces. The collective weight of all individuals of the five rarest butterflies that I discuss in this book would weigh only three pounds five ounces—as much as one panda's paw. And, in contrast to these tiny populations, there are billions of individuals of such common butterflies as Painted Ladies (*Vanessa cardui*) and Small Cabbage Whites (*Pieris rapae*).

The rarest butterflies have not always been rare. Some were very abundant until the last few decades; it is likely that their numbers dropped from millions to thousands. For other rare butterflies, it is impossible to estimate their historical abundance. However, we do know the historic range of their habitats and from that we can extrapolate high abundances. Global habitat loss and climate change have relegated each species to minuscule land parcels, areas as small as a single golf course or even a football field. I have found rare butterflies in unexpected places, their populations restricted to artillery ranges or beaches or backyards.

In every year that I worked and for every species that I studied, I wondered whether I would see the last of these butterflies. The rarest butterflies fly dangerously close to extinction. Numbers are so low that I feared small changes in the area of a forest, the saturation of a wetland, or the level of the ocean would wipe out an entire species. Their numbers and ranges are so small and the threats are so high that my encounter with the last butterfly was a real possibility.

I have staked much of my professional career on efforts to reverse butterfly population declines. There remains some glimmer of hope in prospects for species recovery. In this book, I recount stories about my and others' progress in understanding the biology and conservation of the rarest butterflies in the world. Moreover, I argue that they stand among the poster animals for the loss of biodiversity and the future of conservation. The rarest butterflies may seem at first a surprising or even undeserving part of this group. They are a small set of virtually unknown animals that may appear more idiosyncratic than emblematic of environmental biology and conservation. Yet, viewed in a deeper way, the rarest butterflies provide a unique lens into growing concerns and problems: the loss of biodiversity on our planet and the challenges associated with conserving species in peril.

BUTTERFLIES AND GLOBAL CHANGE

The rarest butterflies suffer from headline-grabbing environmental catastrophes, such as habitat loss, climate change, environmental toxins, and invasive species. Often, these threats act in concert to reduce their populations. The rarest butterflies—and, I argue, much of the diversity of life on earth—are confronted simultaneously with multiple threats that accelerate decline.

Even when habitats are viewed as protected, they can be lost from the perspective of the rarest butterflies. This is a key to understanding why the rarest butterflies are so rare. Like the fairy-tale character Goldilocks, the rarest butterflies require conditions that are *just right*. Some butterflies live in habitats maintained naturally by disturbance such as fire. Too much fire over too broad a region will incinerate populations. Too little fire will cause butterfly habitat to disappear through natural processes of succession, causing their host plant (defined as the plant or plants that caterpillars live on and eat) to die. By stopping fire, draining wetlands, and stabilizing beaches, people suspend natural environmental change and upset a delicate balance. Conditions are no longer *just right*. These insidious threats, cast against a backdrop of major global changes, are slowly eroding the populations that remain.

One thing that distinguishes the rare butterflies from other rare species is that they have specialized environmental requirements that exist in places where people also want to be. In some cases, people use land in ways that are incompatible with the habitat required by the butterflies. Some butterfly ranges have the bad luck of being located near dense urban development or large monocrop fields. In other cases, the expanding footprint of people has helped to conserve the rarest butterflies. The St. Francis' Satyr, for example, lives only on a mostly undeveloped army base. The message here is that there can be win-win scenarios for people and butterflies.

THE SCOPE OF MY SEARCH

Is it even possible to identify *the single rarest butterfly* in the world? As I attempted to do so, I wrestled with issues that arise in the conservation of all plants and animals about how to define what is rare. Many, many butterflies are rare, a number that is too great to cover in many book volumes, let alone in one book. The growing number of rare butterflies is an inevitable consequence of global environmental change. I narrowed my scope to those that I found to be the rarest. As I will discuss, I consulted references worldwide about butterfly population sizes. I failed to identify a species outside of the United States that, after considering range-wide population estimates, was rarer than the rarest species I identified within the United States. In part, the list also reflects my personal journey, restricted mainly to North America. Although others might dispute my assessment and ranking of the rarest butterflies, the conservation needs of each species I describe are not in doubt.

To guide my search, I drew heavily on lists of butterfly species that have garnered formal recognition as conservation priorities. As people have increased their attention to butterflies and their rarity, political processes have evolved to favor butterfly protection. As I looked back at the history of butterfly conservation, I saw plainly that the enactment of the US Endangered Species Act (ESA) in 1973 was a watershed. The first butterflies appeared on this list in 1976. Species are recognized, or listed, as either *endangered* (in imminent danger of extinction) or *threatened* (in danger of becoming endangered).

While the Endangered Species Act applies to the United States, the International Union for Conservation of Nature (IUCN) maintains the most prominent worldwide list of species that have disappeared or are threatened with extinction: the IUCN Red List of Threatened Species. The IUCN began to document the conservation status of butterflies in 1983. I was attracted to the Red List because it adopted *quantitative* scales of vulnerability that included population size, range, and change over time. Another international list was compiled, beginning in 1976, pursuant to the Convention on International Trade in Endangered Species (CITES), an international agreement designed to protect threatened species, including rare butterflies. This list identified rare species in most danger of being moved by people across national borders; prominent species on the list include tigers and rhinos, the hides, bones, and/or horns of which can be transported across borders. Although less attention is given to butterflies, CITES recognizes species that might be the targets of butterfly collectors.

I also consulted results of efforts to monitor butterfly populations, especially over the area of states or nations. The world's most intense and longest-running butterfly monitoring program is the United Kingdom Butterfly Monitoring Scheme (UKBMS). Others include South Africa's intensive records of long-term diversity and status of butterflies, and the annual butterfly counts of the North American Butterfly Association (NABA). Even after I consulted these resources, I recognized that there were surely other rare butterflies awaiting discovery. Those were beyond my purview for this book.

I used these resources to narrow the list of the world's rarest butterflies. I gave a great deal of thought into how to assess rarity. Is the rarest butterfly the one with the smallest number of individuals that remain? Well-established scientific theory and observation have shown that biological and genetic factors can drive numbers in small populations down further. Or is the rarest butterfly the species that ranges globally over the smallest area, measured in the tens of acres and often in remote locations? A small area exposes rare species to large and rapid decline. Long-term changes, such as habitat conversion to cities or fields, and short-term changes to the physical environment, such as drought, can change a butterfly's environment quickly throughout its range. Perhaps the rarest butterfly species is the one whose population is experiencing the most precipitous decline. Some of the rarest butterflies flew over large regions until recently, areas the size of half a state or province. In some instances, scientists ignored them until populations declined to the last butterfly. Should I have considered how novel a butterfly is in the context of all butterflies and of life on earth? A lone and unique lineage may have greater value to genetic diversity and future evolution.

As there was no single standard, I chose as my criterion the total number of individuals of the species left in the world. This measure was the most transparent and the most directly linked to conservation threat. When I began my search, I expected that I could find a scientific study reporting total population sizes for each butterfly species purportedly among the rarest. In practice, scientists adopted a variety of methods to derive butterfly population numbers. For one example, scientists counted the number of Eastern North American Monarchs (*Danaus plexippus plexippus*) on a few trees in the butterflies' overwintering grounds in Mexico and then applied that density to the few hectares they cover there. Others caught the world's remaining Schaus' Swallowtail females (one measure) and propagated them in a lab for release into the wild (another measure). One thing I enjoyed about writing this book was the opportunity to distill the results of different methods to numbers that I could compare.

> ## BUTTERFLY LIFE CYCLE
>
> A butterfly passes through several stages as it completes its life cycle from egg to adult. Caterpillars (also called larvae) grow in about five stages; each stage is called an instar. When an instar outgrows its skin, or exoskeleton, it sheds it (molts) and passes to the next instar. After its final molt, it is left with a hardened outer shell, or chrysalis. The chrysalis contains the inactive larval form (a pupa) that transforms into an adult via metamorphosis. Butterflies can live through one or more generations each year. They can enter an inactive state, or diapause, to avoid harsh conditions such as winter weather. Different species diapause at different life stages, and the length of time a species spends at any stage is variable. For example, whereas adult, overwintering Monarchs can live for six months, adult St. Francis' Satyrs live for only four days.

In this book, I tell the stories of eight species and subspecies of butterflies. I have organized the book by chapter, covering the six rare butterflies in sequence from the most common (Bay Checkerspot, *Euphydryas editha bayensis*) to the rarest (Schaus' Swallowtail, *Heraclides aristodemus ponceanus*); following these are chapters on an extinct subspecies and a common subspecies. Figure 1.1A shows the large range in numbers across all eight species and subspecies. For a given species, all of the butterflies might or might not have lived in one place. To a conservation biologist, a population encompasses the area in which all the individuals of a species can interact with one another. One population is geographically separated from others. As a general rule, the total number of individual butterflies correlates with the number of populations. For example, the Fender's Blue (*Icaricia icarioides fenderi*) occurs in more populations than and at higher abundances than the Schaus' Swallowtail. The situation is different with the Bay Checkerspot and the Eastern North American Monarch, which are the most abundant butterflies I discuss, but which live in one population (Figure 1.1B). Among the eight butterflies I looked into, the number of populations ranged from one to thirty-six, and my search turned up some butterflies that occupied few places but were, in fact, very numerous relative to the others.

SUBSPECIES

Scientists have classified most of the butterflies featured in this book not as species but as subspecies. A species, by definition, is the group of individuals that can interbreed only with one another. The subspecies category recognizes that some individuals of the same species are so geographically isolated—by, for example, distance or mountain barriers—that they will never have the opportunity to interbreed. Unlike separate species, members of different subspecies could, if brought together, interbreed. Because they are separated by such large distances, different subspecies may differ in their color, form, or behavior.

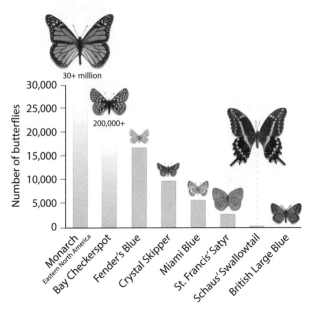

FIGURE 1.1A. Worldwide number of butterflies of each species or subspecies highlighted in this book. I explain the data and give sources in each chapter.

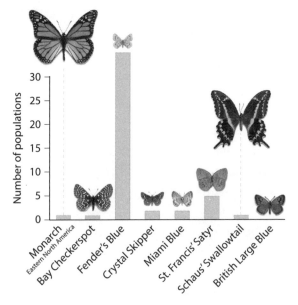

FIGURE 1.1B. Worldwide number of populations or metapopulations of each species or subspecies of butterfly. I explain the data and give sources in each chapter.

One butterfly I studied in North Carolina, the St. Francis' Satyr, illustrates the challenges and consequences of recognizing subspecies. At first appearance, the St. Francis' Satyr looks identical to a related subspecies, the Mitchell's Satyr (*Neonympha mitchellii mitchellii*), which was until recently found only in Michigan (small populations have turned up in Alabama, Mississippi, and Virginia). The scientists who first discovered the St. Francis' Satyr named it a subspecies because of its distance from the other subspecies and subtle differences in appearance. Does this designation hold up when evaluated by other standards? To learn more, I collaborated with a graduate student, Chris Hamm, who gathered samples and tested how closely related the populations are to one another. Chris conducted genetic analyses of six short segments of DNA. His analysis revealed that the Mitchell's Satyr and the St. Francis' Satyr are different, perhaps even different species.

I included subspecies in my search because they are significant in conservation work. Subspecies are important in four ways. First, subspecies and not species represent more fully the variation of life on earth. Second, a subspecies' decline in a region is indicative of broader environmental degradation. Third, subspecies play an irreplaceable role in ecological systems. Fourth, one subspecies can provide a source to replace another subspecies of the same species if one goes extinct.

For example, one subspecies of Large Blue butterfly that occurred in England, the British Large Blue (*Maculinea arion eutyphron*), went extinct before scientists understood how to conserve it. With new knowledge in hand, they were able to introduce another subspecies of Large Blue (*Maculinea arion arion*), from Sweden. It was successfully established. Although the genetic variation carried in the British subspecies was lost forever, the Swedish subspecies could for the most part fill the

British Large Blue's ecological role. With this in mind, I chose to focus on subspecies because of their need for conservation and their relevance to ecological systems. For the remainder of the book, I sometimes use the term *species* for simplicity, whether I am talking about species or subspecies.

FRAGILE OR RESILIENT?

Despite the threats faced by the rarest butterflies, these insects persist. Their persistence embodies a prominent question in conservation: Is nature fragile or resilient in response to environmental change? Some see resilience in nature, including in the recovery of some large animals and in the apparent functioning of ecosystems after elimination of some animal and plant species. Those who see fragility can point to mounting extinctions of birds, mammals, and other wildlife, and broader degradation of ecological systems. The rarest butterflies provide an intriguing lens through which to examine this dichotomy.

Their delicate wings and small size suggest fragility. I often use the most important item of research equipment in my toolbox, a butterfly net, to brush over vegetation and instigate butterfly flight. At times I use my net to capture butterflies, so I can mark their wings and track their populations, or to catch females that will lay eggs for captive propagation, or to sample tiny wing fragments for genetic analyses. When I do this carefully, I find that individual butterflies are resilient to capture and handling. If I capture them improperly, however, I can loosen scales or bend wings. Their wings mirror a fragility in natural environments that is a more urgent threat to their survival. All the rarest butterflies have declined and are declining rapidly. Each is slipping away.

Despite their fragility, I am inspired in my research and conservation by signs of resilience. I am heartened by the possibility of restoration and recovery. One of these signs is the persistence of many of the rarest butterflies in unnatural places. A theme that repeats across chapters is that activities of people can sometimes replace natural disturbances (fire, for example) that they have otherwise reduced or eliminated in butterfly habitat.

There is not a strict dichotomy between fragility and resilience of the rarest butterflies. I find compelling a story of William Henry Edwards, a nineteenth-century lepidopterist (a person who studies or collects butterflies or moths, which belong to the scientific order Lepidoptera). He lived along the Kanawha River in West Virginia. A swamp near Edwards's home had a healthy population of Baltimore Checkerspot (*Euphydryas phaeton*) butterflies. Local coal companies sought to improve river transportation, and to do so they made waterways deeper and wider and created a series of locks and dams that flooded the landscape. Consequently, they exterminated the fragile population of the Baltimore Checkerspot. Keen to restore this butterfly's population, Edwards waited until the floods receded. He then propagated Baltimore Checkerspots from caterpillars to adults and released them back into the river valley. The population flourished.

I find that the rarest butterflies embody a more fundamental dichotomy between the gloom and hope that I share with others involved in modern conservation efforts. On the front lines of conservation, the signs of decline among rare butterflies are everywhere, and it is easy to despair. Yet, as a conservation biologist, I am optimistic and hopeful that the demands of people can be reconciled with the needs of nature. The rarest butterflies have taught me how people are causing environmental harm, but my research has also led to discoveries that expand possibilities for healing nature.

Even if resilient, the rarest butterflies will not recover immediately. It grates on me when I read opinions written by politicians or skeptical conservationists who see failure when a butterfly that receives legal protection as endangered, accompanied by investment of time and money in conservation, does not recover in five years or a decade. The rarest butterflies have arrived near extinction over decades or centuries of decline. Is it then surprising that it could take just as long to see them recover? By unraveling their biology, I hope to learn enough about the rarest butterflies and their ecosystems to restore them. In this book, I contemplate the fragility and resilience of nature as it tells the story of the loss and potential recovery of the rarest butterflies.

AVERTING EXTINCTION

The most basic goal of my research is to pull the rarest butterflies back from the brink of extinction. There are metrics other than extinction for measuring the effects of changing environments on butterflies. However, once they have fallen to ultimate extinction, conservation is irrelevant to them, and people have reduced earth's biodiversity. As I tell the stories of individual rare butterfly species, I will be relating what scientists are learning about recovering their populations before they reach this irreversible fate. Yet, for each of the rarest butterfly species in the world, extinction remains a very real possibility.

As far as we know, only three butterfly species and a dozen butterfly subspecies have ever gone extinct. This apparently low number might be because butterflies have weathered global environmental change. More likely it is because undiscovered butterflies have gone extinct without notice. One butterfly species that went extinct was the Xerces Blue (*Glaucopsyche xerces*). It inhabited sand dunes in what is now San Francisco's Sunset District. The gold rush that began in the late 1840s caused the city to fill with people. By 1875, biologists recognized the Xerces Blue's decline. Herman Behr, a curator of entomology at the California Academy of Sciences, wrote, "The locality where it used to be found is converted into building lots, and between German chickens and Irish hogs no insect can exist besides louse and flea." By 1941 the Xerces Blue was extinct.

Now the only place to see the Xerces Blue is in museum collections. My search for it took me to the McGuire Center for Lepidoptera and Biodiversity at the University of Florida in Gainesville. I asked collections manager Andy Warren to show me a pinned specimen of the Xerces Blue. I gasped when he pulled out a drawer containing more than one hundred specimens collected seventy-five or more years ago. There was a time in some people's living memory when Xerces Blue was not so rare.

Scientists presume two other butterfly species are extinct. Both occurred in South Africa. The Mbashe River Buff (*Deloneura immaculata*) was collected only three times ever, all in the mid-nineteenth century on the Eastern Cape. Morant's Blue (*Lepidochrysops hypopolia*) was also collected three times, in the 1870s, in two different regions. Nobody has observed a single individual of either since then.

I include a chapter on one extinct subspecies, the British Large Blue. Beginning in the nineteenth century, it declined in abundance for over one hundred years. Scientists learned the key details of the butterfly's biology and conservation just as it went extinct. The British Large Blue exemplifies how declines and recovery of the rarest butterflies require understanding of the very subtle and particular biology of butterflies. I include this chapter because it carries this and other lessons about the conservation and science of the rare butterflies that still exist.

THE RAREST BUTTERFLIES AND BEYOND

Throughout my search, three questions haunted me. First, for each butterfly: What can I or others do to reverse course and prevent impending extinction? Second, stepping back from each butterfly: Can general lessons learned from one or a few butterflies be applied to other butterflies and other animals and plants? Third, the hardest question of all: Are there compelling reasons to save the rarest butterflies in the world—or should we put our efforts elsewhere?

I have discovered that the rarest butterflies in the world are emblematic of the consequences of a range of global environmental changes and of the modern challenges in biodiversity conservation more generally. Throughout the book, I contemplate and clarify the value of species and the meaning of their potential loss. By stringing together observations that connect biology to global change to conservation, I have come to know with more intimacy the diversity of life on earth and its need for protection.

Although my focus is on very rare butterflies, even the most common species are susceptible to precipitous population declines. For this reason, I have included a chapter about the Monarch. It shares with the rarest butterflies threats from all environmental changes, and because of this, it too is becoming rarer.

I will admit that I included the Monarch with some reluctance. Anyone reading this book will know the Monarch. For many people, any familiarity with butterfly biology is limited to a knowledge of Monarchs. When I have given talks about rare butterflies, I have fielded questions about why these species cannot do what Monarchs do. I once talked about the destruction of a rare butterfly's habitat. A student asked, "Why don't the rarest butterflies just migrate to better places, far away from danger?" In their migration and in other ways, Monarchs are unique, and they do not provide a good model for other butterflies. Unlike Monarchs, which are widely distributed and familiar, the rarest butterflies are unknown to nearly everyone. Still, the Monarch is declining. Scientists are working to understand the causes, so that the Monarch doesn't end up as one of the rarest butterflies. Perhaps one day the knowledge others and I acquire about recovery of the rarest butterflies will inform conservation of now common but declining species such as Monarchs.

In contrast to the prospects for the rarest butterflies in this book, the idea of the last butterfly in the global sense seems preposterous. Or I thought so, until I read an important paper by Stanford University professor Rodolfo Dirzo and colleagues. These researchers reviewed the world's half-century-long records of insect abundances and discovered that butterflies and moths as a group lost on average a third of their numbers (Figure 1.2). This includes measures of common and rare species. At this rate, the last butterfly would be only a few decades off. Clearly, this trend will not continue downward until 100 percent of all butterflies are lost. However, it speaks to the sustained loss of many rare and common butterflies. I was attracted to this paper because of its inclusion of butterflies. As I studied the article's key graphic, I was astonished to learn that all other insects had declined even faster than butterflies. The study of individual butterfly species has such a rich and extensive history that, as I suspected, the analysis included many more members

18 CHAPTER 1

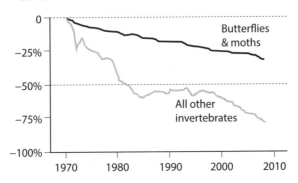

FIGURE 1.2. Change in abundance of invertebrates worldwide since 1970. Adapted from Dirzo et al. 2014, who conducted a review of all long-term studies of invertebrate populations.

of the order Lepidoptera than all other insects combined. It follows that the correlation between butterflies and other insects supports the use of butterflies as a broader indicator of biodiversity decline. Global analyses like this one indicate that the rare and declining butterflies serve as the new canaries in the coal mine for biodiversity loss.

By 1970, the starting point for records studied by Dirzo et al., the rarest butterflies I write about had already traveled downward beyond the low points shown on the graph (Figure 1.2) of 30 percent loss for all butterflies (common or rare) and 80 percent loss for other insects. Now that we have arrived at this point, what can be done? For the rest of the book, I try to answer this question for each of the rarest butterflies, highlighting new advances in science and conservation. Perhaps these answers can be used to halt or reverse the downward trend for the rarest butterflies and, ultimately, for other insect species.

PART I
THE RAREST BUTTERFLIES

CHAPTER 2

BAY CHECKERSPOT

When I lived in the San Francisco Bay Area in college, I was oblivious to the region's status as a global epicenter for rare butterflies. The Mission Blue (*Icaricia icarioides missionensis*), the San Bruno Elfin (*Callophrys mossii bayensis*), and the Callippe Silverspot (*Speyeria callippe callippe*) are all found in and adjacent to the two-thousand-acre San Bruno Mountain State Park, located on the southern boundary of San Francisco. Other rare butterflies in the region include the Lotis Blue (*Lycaeides idas lotis*), the Behren's Silverspot (*Speyeria zerene behrensii*), and the Myrtle's Silverspot (*Speyeria zerene myrtleae*) to the north; the Bay Checkerspot (*Euphydryas editha bayensis*) to the south; and the Lange's Metalmark (*Apodemia mormo langei*) to the east. In all, eight butterflies living within a sixty-mile radius of San Francisco are listed on the US Fish and Wildlife Service's threatened and endangered species list. Of the twenty-six threatened or endangered butterflies in the United States, one of every three is found in this small region.

The Bay Checkerspot, a butterfly with a two-inch wingspan and wings covered with rows of orange and white or yellow spots against a black background (Plate 2), is one of the Bay Area's rare butterflies. Of all the world's rare butterflies, the Bay Checkerspot is one of the two most studied species (the other being the Large Blue in England). It was with this butterfly that Stanford University professor Paul Ehrlich conducted his classic ecological studies in population biology. His pioneering work resulted in some of the precursors to modern spatial ecology and conservation biology, disciplines that are key in the development of principles regarding how to manage populations in landscapes fragmented by people. Paul's sustained research on the Bay Checkerspot led to a series of discoveries by his lab and collaborators that are now propelling the butterfly's conservation.

As an undergraduate, I worked in Paul's lab but not on butterflies. After failed starts in engineering and then economics during my first two years of college, I'd finally set my career path in conservation biology. I had no vision of what I would do in this field. At minimum, it would take me outdoors to conduct science. On my inaugural research trip, I remember walking across an open, grassy hill at Stanford's Jasper Ridge Biological Preserve. I did not know it at the time, but this trip brought me within just feet of the Bay Checkerspot.

In 1990, I headed into the field to study birds as an assistant to one of Paul's graduate students, Tom Sisk (now a professor at Northern Arizona University). I studied patterns of birds across the varied habitats of the California foothills, including oak woodlands, shrub-dominated chaparral, and grasslands. As I traversed the grassy ridge, I passed another student doing field research. It was not until months later that I learned that the student was studying the Bay Checkerspot. Even more surprising, my path wound through three sites that had supported a historic population of the butterfly. Unfortunately, the population at one of those sites had already been lost, and the other two were set on a downward trend. By 1998, the population at Jasper Ridge was extirpated (a term I use to differentiate the loss of a population from extinction, or the loss of an entire species or subspecies).

TWO CENTURIES OF DECLINE

The extirpation of the Bay Checkerspot population at Jasper Ridge was not the first. Over the previous decades, other populations nearby had also been lost. The forces of environmental change that caused the Bay Checkerspot (and other butterfly species living nearby) to become perilously rare were set in motion long before San Francisco's rapid growth. As far as we know, the Bay Checkerspot lived in large areas that people have never converted to cities or farm fields. No one intended to destroy the Bay Checkerspot's habitat. Yet, the population growth in the Bay Area led to large, unintentional changes that caused the Bay Checkerspot to decline. The sequence of change unfolded over two centuries.

The change began when Spanish settlers arrived in the Bay Area. In 1776 they established permanent occupancy, and an immigrant population soon gathered size and strength. In 1824, the Spanish missions began dividing their land holdings into separate parcels, ushering in a dramatic change in land use. Change in land ownership accompanied an increase in one particular use of this land: cattle ranching. By the 1830s, the Bay Area harbored hundreds of thousands of grazing cattle, horses, and sheep. Livestock farming transformed the landscape in two ways. First, the animals consumed and trampled native plants. Second, the farming practices spread populations of invasive grasses, which came to dominate the region.

Invasive species caused immediate change and set the stage for future habitat degradation. The Bay Checkerspot lives in grasslands that cover the foothills that encircle the bay. At first glance, it looks as if there should be plenty of habitat for the butterfly; however, not all grasslands are equal. Native California grasses and herbs, including the host plants with which the butterfly evolved, no longer dominate vegetation in the foothills. Invasive Italian Ryegrass (*Lolium multiflorum*), Common Wild Oat (*Avena fatua*), Soft Brome (*Bromus hordeaceus*), and other species do. These species grow fast, produce many seeds, and establish easily. These and other characteristics allow them to displace native species. Now the butterfly's host plants are relegated to a few small habitat patches scattered across the Bay Area.

The landscape's degradation continued apace with the gold rush of the 1840s and 1850s. Houses and buildings covered some of the Bay Checkerspot's habitats. More important, urban expansion near the species' dwindling habitat caused unusual changes that proved the most enduring cause of butterfly decline.

DISCOVERY AND DISTRIBUTION

By the early twentieth century, most of the Bay Checkerspot's habitat had been lost to invasive species, and the butterfly persisted in tiny fragments of remaining habitat. By the time Robert Sternitzky of the Pacific Coast Biological Service discovered the Bay Checkerspot in 1933 (it had been unknown to science before his discovery), the butterfly occupied only a few small areas that were unique in their ability to withstand invasion by exotic plants. The butterflies found a foothold in grasslands that grow on a soil type that is uncommon in the Bay Area (Plate 3, bottom). This soil forms on outcroppings of serpentine rock, which has a greenish hue and slippery feel and comes to or near the surface of the thin soil layer. Serpentine soils have an unusual composition of nutrients. They are low in some nutrients that are essential to living things, such as nitrogen and calcium, and high in other elements that can be toxic, such as magnesium and nickel. Besides its unique blend of nutrients, serpentine soil dries quickly. These features of the environment support a unique community of plant species. The level of soil nutrients is an important part of the story that I will come back to, as the actions of people are changing natural levels of these nutrients and putting the Bay Checkerspot in greater jeopardy.

Bay Checkerspots lay most of their eggs on Dwarf Plantain (*Plantago erecta*). This host plant for the species' caterpillars is a small herb that grows to about one foot in height, often in serpentine soils. Its narrow leaves cluster around the plant base. Other aspects of the butterfly's biology besides its selective diet enable it to persist in serpentine habitats. Unlike most butterflies, which lay their eggs individually or in small clusters, Bay Checkerspot females dump large masses of up to 250 eggs on single, small plants.

After the Bay Checkerspot's eggs hatch, tiny young caterpillars form webs and live communally. Dining in concert, they are able to ravish their diminutive host plant. Unlike many caterpillars, which can feed on the plant on which they hatch until they become a chrysalis, Bay Checkerspot caterpillars exhaust their first host and must crawl to other plants nearby. When they set out, their destination is not only another Dwarf Plantain. The Bay Checkerspot has two other host plants, Purple

Owl's Clover (*Castilleja exserta*) and Denseflower Indian Paintbrush (*Castilleja densiflora*). The three host plants grow in different seasons. The Bay Area's winter rains are followed by eight arid months. In summer or fall when I scan the grass-covered foothills, the only color I see is brown. As the landscape dries out, Dwarf Plantain desiccates. It can do so before the caterpillars have finished eating (Plate 3, top). If it does, the caterpillars must search for a new host plant. Purple Owl's Clover and Denseflower Indian Paintbrush emerge later in the wet season and remain green for a short time into the dry periods. The caterpillars' goal is to make it through about half their life cycle before all of their host plants dry out. When they have made it this long, they retreat under dead leaves or rocks, where they enter diapause for seven months. After the rain returns, their host plants form new leaves, and the Bay Checkerspots can complete their life cycle. In a good year, about half the caterpillars that live through diapause will complete their life cycle. In a bad year, for example when plants are under stress from extreme drought, only a quarter may survive this gauntlet.

Serpentine soils provide safe havens for Bay Checkerspot populations; however, the dozen populations that remained on these soils in the middle of the twentieth century were not safe. They ranged over fifty miles, from just south of San Francisco to just south of San José. Some areas where they thrived in and near San Francisco, such as Twin Peaks, became part of the city, covered over with homes and other buildings. A large freeway sliced into Edgewood Park, just south of San Francisco, where another Bay Checkerspot population occupied about 115 acres until 2002. The now-extirpated population at Jasper Ridge once inhabited a small grassland covering just twenty-five acres. The best conservation outcome would see Bay Checkerspot reestablished in the minuscule number of areas distinguished by serpentine soils that remain undeveloped.

POLLUTION OF THE LAST REFUGE

By 2002, amid this ongoing loss, a single, large population of the Bay Checkerspot remained. Coyote Ridge (Plate 3, bottom) extends over an area of five-thousand-plus acres at the southern edge of the butterfly's historic range. Because of its outlying location, Coyote Ridge evaded the development that has engulfed the ring of cities surrounding the bay. By nearly all measures, Coyote Ridge appears in excellent condition to support a large population of Bay Checkerspots.

Even in this large and reasonably protected environment, threats persist. Although the habitat appears untouched, people are nonetheless changing Coyote Ridge's serpentine grasslands. Indirectly, they foster establishment and growth of harmful invasive species. The origin of this change is pollution by nitrogen, an element that is naturally rare in serpentine grasslands. People and their activities can tip the balance to favor nitrogen, which is both a fertilizer added to soils to feed plants and a pollutant. Particularly bad sources are automobiles. Each automobile emits oxides of nitrogen during combustion. Vehicle exhaust travels through the air before falling onto grasslands. The closer the habitat is to a highway, the more of this airborne pollution, including nitrogen, it receives.

Stuart Weiss of Creekside Center for Earth Observation discovered the effects of nitrogen pollution on butterfly loss. Stu has been the Bay Checkerspot's steward for three decades. Among his other studies, he analyzed nitrogen levels recorded at long-term monitoring stations and simultaneously at several current and former butterfly habitats. He found that nitrogen deposition in the Bay Checkerspot's habitats is now between five and fifteen times higher than historical background rates. These elevated levels of nitrogen feed invasive species and degrade butterfly habitat. In theory, Bay Checkerspot restoration needs reduced air pollution. Given the high number of people and automobiles in the region, it seems an unimaginable goal to reduce levels low enough to sustain native grasslands. One can only hope that electric cars will replace internal combustion vehicles in the nick of time.

While it may not be possible to prevent nitrogen from infiltrating native grasslands, it may be possible to control the fundamental threat to the Bay Checkerspot, invasive grasses. To control them, Stu's group turned to unconventional means. He found that cattle, an agent of plant invasion, have become an unexpected ally. Where cows feed on serpentine soils, their menu focuses on tastier, non-native grasses. In 1996, Stu tested the hypothesis that cattle removal caused butterfly decline. He established study plots within a Bay Checkerspot population in which he either allowed or excluded grazing. He found that when sites were closed to grazing for five years, the abundance of the butterfly's host plants declined by half. Grazing maintained the balance in favor of native plant species and therefore in favor of the Bay Checkerspot.

Accordingly, cattle create a dilemma for the Bay Checkerspot. Too many cattle will disturb the soil through trampling and alter nutrient cycles through their excrement. The conservation challenge is to find just the right balance of grazing cows or other unnatural disturbances to manage grasslands for butterflies. This general challenge reappears with different agents of disturbance in other stories of the rarest butterflies. An uncomfortable realization for me was that disturbances can actually help to maintain native grasslands for butterflies.

CLIMATE VARIATION, NATURAL AND UNNATURAL

At Coyote Ridge and elsewhere, another factor that affects Bay Checkerspot populations is climate variation. Even in the absence of people, Bay Checkerspots live in a variable climate. The life cycle of the Bay Checkerspot requires the right amount of rain to sustain caterpillars. California's climate is like other Mediterranean climates in that, throughout, precipitation varies between wet and dry seasons. Extreme variation results in either surplus rain or drought. In the hottest, driest years, host plants will dry out before caterpillars reach milestone sizes in their yearly life cycle. To protect against possible change, female Bay Checkerspots are so-called gamblers, choosing between two strategies. They can lay fewer eggs early, giving their caterpillars a chance to feed and survive through the season. Or they can roll the dice and lay more eggs later, when the race against time becomes accelerated. As a rule, female butterflies opt for more eggs later, and their offspring are left to race against time.

We can predict the potential consequences of drought on the Bay Checkerspot based on historical data from the extirpated population at Jasper Ridge. Four decades of data reveal a span from very high to very low amounts of rainfall. One period had particularly strong effects on Bay Checkerspot populations. Rainfall at half of average levels in winter 1975–76 preceded rainfall at three-quarters of average levels in winter 1976–77. Paul Ehrlich and his collaborators found that these successive droughts reduced plant growth and the duration of green leaves. Butterfly populations plummeted to about one-fifth their pre-drought levels. To put that in perspective, think of the people in your community and then remove four out of every five of them.

Drought is not the only change in precipitation that causes Bay Checkerspot populations to diminish. A deluge of rain can be just as bad. In rainy years, many caterpillars lose the race against time. High levels of precipitation correspond to cloudy and cold conditions that impede the growth and development of the caterpillars. They enter the dry season with stunted growth, and after the leaves on their host plants turn brown, the undersize caterpillars perish.

While Bay Checkerspot's populations fluctuate naturally in response to variable precipitation and temperature, their sensitivity makes them especially vulnerable to the effects of climate change. Butterflies could weather declines that were not too severe or frequent as long as they were able to recover in moderate years. Climate change is producing more excessively wet years and excessively dry years. When these extremes come in succession, they can be devastating.

The rolling landscapes where Bay Checkerspots live buffer them against extreme climate fluctuations. Temperature varies across these hilly landscapes, and these varying conditions can occur contiguously in the rolling terrain. Since temperature affects the ability of caterpillars to grow (they grow faster in hotter places) and of plants to remain green (they turn brown in hotter, drier environments), the varying temperatures give Bay Checkerspots a range of options for adjusting to climate fluctuations. Stu measured air temperature and caterpillar temperature, growth rate, and dispersal on hills with different slopes and orientations. He found that north-facing slopes were cooler, south-facing slopes were hotter, and flat hilltops were in between. This means that in the same small area, one hillside exposes caterpillars to a different temperature than others. These temperature ranges can be large, as much as 27 degrees Fahrenheit. Faced with this range, the Bay Checkerspot must find the right balance between growing too fast and too slow in places where there will be more or less food. Caterpillars can crawl tens of meters between one slope and another. Complex topographic environments, like those found at Coyote Ridge, provide the best conditions for Bay Checkerspots that live in a variable and changing climate.

THE ROAD TO RECOVERY

In 1994, I visited Coyote Ridge and finally saw the butterfly that had eluded me at Jasper Ridge. I was surprised by Coyote Ridge's large area and encouraged by the commitment to enlarge it further. Restoration of the Bay Checkerspot's habitat at Coyote Ridge had focused on removing invasive species and reintroducing disturbance via limited cattle grazing. Support for restoration had come in the form of a government-sanctioned habitat conservation plan and funding meant to reduce the potential environmental damage caused by adjacent centers of energy, transportation, and waste disposal. Stu told me that now when he walks through this area during the spring, when flowers are blooming and butterflies are flying, "it feels like I am walking through a Monet painting day after day." In reference to the butterfly's host plant, he described the study sites as "plantago-licious."

Stu's efforts at Coyote Ridge provide a template for Bay Checkerspot recovery. Beyond that, I look to his work as a model for my research on and conservation of other butterflies. In 2015, Stu's group estimated that there were nearly two million Bay Checkerspot caterpillars. About half of those caterpillars died in development. All told, the estimate of the adult butterfly population size was one million. This size provided a buffer against natural fluctuation. After three successive years of bad weather starting in winter 2016, for example, the population size had declined to about one hundred thousand. In good years, the population size should rise again. Considering this range, the number of these butterflies is large, especially for a subspecies classified as threatened.

Despite its large size, the Coyote Ridge population is only the start of recovery. A major concern is that there is not a second (or third or fourth) population. There are two ways that more populations are important. First, another population would serve as a safety net in the event of extirpation at Coyote Ridge. Second, a series of populations would mimic the natural dynamics of Bay Checkerspot populations. The Bay Checkerspot once formed a population of populations. Each occurred on one of a series of serpentine grasslands that were scattered across the landscape. This is a classic example of what ecologists refer to as a "metapopulation" (Figure 2.1). Habitats can vary in quality among populations over time. This may be due to human-caused threats that are more severe in one habitat than in another or to natural causes as different populations experience different environments, with higher or lower elevation, north- or south-facing slopes, or more or less rainfall. Natural or not, variability in environmental conditions causes populations to increase and to decline, and in the worst case a small population will be extirpated. Historically, butterflies occupied habitats close enough to one another that movement between them balanced the loss of any individual population. The process of reestablishment is a feature of a stable metapopulation.

Unfortunately, habitat fragmentation in the last half century confines Bay Checkerspots to the places where they are born. Coyote Ridge is at too far a distance to provide immigrants to repopulate areas where populations have been extirpated.

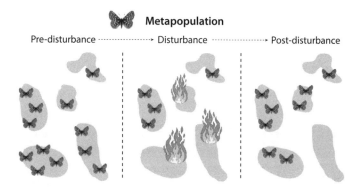

FIGURE 2.1. A metapopulation is a set of separate populations, in which each population is within target distance of dispersal to another population. The natural dynamics of a metapopulation include extirpation of a population (for example, after disturbance) and immigration to replace extirpated populations.

Cities, suburbs, and foothills separate habitats by tens of miles. The Bay Checkerspot has very few options for recovery. At this stage, any realistic attempt to rescue it will require ongoing efforts by people. In addition to protecting and restoring its habitat, people will need to intervene and replace the natural dispersal process by transporting Bay Checkerspots to places where populations once thrived. These two components— restoration and transportation—form the most promising efforts on behalf of Bay Checkerspot conservation.

Stu and other conservationists recognized the critical need to establish new populations outside of Coyote Ridge. His first efforts focused on the site of a historic population at Edgewood Park. I had a chance to tour Edgewood Park with Stu in the spring of 2014 and could see firsthand why the park seemed the ideal target for restoration. It covered nearly five hundred acres of grasslands and woods. Until 2002, Edgewood Park had hosted the last Bay Checkerspot population outside Coyote Ridge. It had outlasted populations at San Bruno Mountain

State Park to the north and at Stanford's Jasper Ridge to the south. It retained many features of high-quality habitat. Serpentine soils dominated forty-five acres of Edgewood Park and supported many of the butterfly's host plants. Despite these features, Edgewood Park needed restoration to reverse habitat degradation.

Before butterfly reintroduction at Edgewood Park could begin, conservation actions were needed to address sources of degradation that had caused the Bay Checkerspot's extirpation there in the first place. Of central importance was a new interstate. The interstate was laid directly over the western part of the park, including some Bay Checkerspot habitat and thus a part of the population. Relative to the size of the entire park, the affected area was small. But the interstate precipitated much larger declines of the Bay Checkerspot.

Automobiles caused a devastating effect. The freeway led nearly one hundred thousand cars past Edgewood Park each day. As vehicle traffic increased, so did the amount of nitrogen emitted by the cars. This, in turn, facilitated more invasive grasses.

As vehicle traffic and nitrogen emissions will not decrease at Edgewood Park, the most direct target of conservation is invasive-plant removal. Stu and his colleagues have developed a three-pronged strategy to accomplish this. First, and most important, each year the park's staff mows two to three acres of the grassland. This simulates grazing disturbance and reduces cover of Italian Ryegrass and other invasive plants. Second, a volunteer team called the Weed Warriors carries out weekly patrols to eliminate, by hand, especially problematic weeds that may harm native plants and butterflies. I observed one of these volunteers during my 2014 visit, and the effort seemed heroic. At the same time, I cringed as I looked at this single person set against a large field of weeds. Third, the group is experimenting with new techniques; the most intriguing to me is the sinister-sounding hydromechanical obliteration. Jets of water vaporize the unwanted plants, leaving their biomass as mulch to feed the ecosystem without disturbing soils.

Having addressed the primary source of degradation to the extent possible, Stu and others developed a vision to bring the Bay Checkerspot back to Edgewood Park. The crux of their plan was to transfer caterpillars from Coyote Ridge. When I compare the Bay Checkerspot to other butterflies I have worked to reintroduce, I am impressed by a distinct advantage: because it lays its eggs in groups that can number into the hundreds, and because the caterpillars initially live together in a colony, it is easy to collect many individuals quickly. Humans can then serve as agents of dispersal, replacing natural pathways that urbanization has severed. Once transported to Edgewood Park, a few individual butterflies can seed a new population.

Stu's first attempts were modest, balancing the benefits of establishing a new population at Edgewood Park against the prospects of slight harm caused by removal of individuals from Coyote Ridge. In 2007, Stu's team transported a few hundred caterpillars. The process was much easier than I could have imagined. With other butterflies, I have painstakingly transferred individual caterpillars onto individual leaves of host plants, assuring myself that they are safe on a plant where they have ample food to get them through their life cycle. Young caterpillars can be small and delicate, and I admit anxiety when risking caterpillar damage or death each time I move one. At Coyote Ridge, Stu's group can collect significant numbers of inch-long caterpillars. I was astonished to learn that after they transport the caterpillars to Edgewood Park, they sprinkle them on the ground as if adding a pinch of salt to a stew. Stu was confident that the caterpillars would have no problem crawling to their host plant.

The group's first efforts followed one of the four deepest droughts in a century, and those initial attempts failed. The next year after the transport and release, Stu's group observed just one surviving caterpillar. I cringe when my own restoration work fails, but I have learned that there can be unanticipated benefits from such efforts. Even unsuccessful restoration and reintroduction efforts serve as experiments, increasing our understanding to improve future attempts.

Stu could overcome one major challenge to reestablishment at Edgewood Park by introducing more butterflies. In doing so, he would be adhering to a central tenet of restoration: sizes must increase quickly. Small populations are at imminent risk of extinction, especially as environments can change in unexpected ways. To overcome these risks, Stu's group received permission from the US Fish and Wildlife Service to increase the number of caterpillars removed from Coyote Ridge and released at Edgewood Park. The number rose from just a few hundred to four thousand individuals, a number still within safe limits. It represents a tiny fraction, four-hundredths of one percent, of the population size at Coyote Ridge.

Larger numbers of caterpillars translated to larger numbers of adults in following years. Stu conducted standardized, yearly counts at Edgewood Park of caterpillars and adults. He found a year-on-year increase from 2011 (120 adult butterflies) to 2014 (800 adult butterflies). This represented the largest number of Bay Checkerspots at Edgewood Park since restoration efforts had begun there. The population size then reversed course, declining to forty-seven adults in 2017. Throughout this period, Stu's group had transported additional caterpillars from Coyote Ridge for release.

Stu explained to me that part of the decline resulted from poor weather conditions. Further, he knows that restoration success depends on population persistence in the absence of new introductions. His group did not release caterpillars in 2018 and still observed forty-three butterflies. Although a small number, it showed that the population could sustain itself for at least one year. But at this population size, the Bay Checkerspot won't persist for long. Larger, persistent populations will depend on fine-tuned and increased restoration.

I have embraced one lesson Stu has drawn from his group's restoration work: it is a mistake to expect that a one-time effort will reestablish a butterfly population or reverse a butterfly's decline. In the case of the Bay Checkerspot, the population in Edgewood Park was lost over decades, and it is unrealistic to expect immediate recovery. Stu's efforts have also taught me that it can take decades just to expose the threats underlying butterfly decline. There may not be a single factor. Rather, multiple global changes can act in synergy to jeopardize populations. Restorationists must be willing to experiment and to learn from successes and failures what measures can be taken to sustain efforts for the long term. Stu says that success will pass the "*O*, for *obvious*, test." After showing signs of fragility for decades, the Bay Checkerspot is finally showing evidence of (human-aided) resilience.

ON THE RIGHT PATH

The Bay Checkerspot has been in decline for my entire lifetime. I have never worked on its restoration directly. However, because I have been affiliated closely with pioneers in the species' ecology and conservation for three decades, it has been formative and has steered my own efforts.

One reason for this is that the Bay Checkerspot faces as many as or more types of threats than any other rare butterfly. In 1987, the US Fish and Wildlife Service recognized it as a threatened subspecies. At the time of its listing, twenty-nine of the Bay Checkerspot's thirty-two known populations had already been lost.

Albeit from a low starting point, growing evidence suggests possible movement toward Bay Checkerspot recovery. There is still a long way to go. With sustained effort and some luck, perhaps successes at Edgewood Park can endure. These efforts will then be a springboard to recovery at other locations in the Bay Area. In 2017, Stu brought his group's restoration techniques northward to the site of a historic population at San Bruno Mountain. In the future, restoration would ideally occur in nearby habitats. This is despite the fact that, because of the intensity of urbanization, a network of populations that form a metapopulation is no longer possible.

For a butterfly that was unknown until fairly recently and has been described as "ordinary" in its behavior and population structure, the Bay Checkerspot has become a source of inspiration for those involved in the ecology and conservation of rare species. Through his work on the Bay Checkerspot, Paul Ehrlich identified the need to focus conservation efforts not just on species but on populations, because, he wrote, "Focusing on the loss of species is focusing on the closing scene of what is ordinarily a long process." In terms of the Bay Checkerspot, part of this long process is the loss of endangered ecosystems on serpentine grasslands in the San Francisco Bay Area. These ecosystems harbor plant species that are now restricted to them, hemmed in by invasive species that cannot tolerate the nutrient-poor soils. These include more than a dozen endangered plant species. In this and in all other cases in this book, it makes no sense to focus on conservation of the butterfly alone. Conservation of the threatened Bay Checkerspot and the endangered grassland species go hand in hand.

Stu recommended that habitat conservation for the Bay Checkerspot focus on a diversity of environments. This focus adds a layer to the dominant paradigm in conservation: to protect larger areas. Topographic variation is a key component of habitat quality, as the steepness and orientation of rolling hills create large climate gradients. Habitat diversity will ensure that parts of each population can endure levels of rainfall that are too high or too low or temperatures that are too high or too low. Larger areas are important, but they will matter for the Bay Checkerspot only if they also encompass a variety of habitats within the landscape. Conservation of a diversity of environments will buffer the Bay Checkerspot from the effects of climate change.

Is the Bay Checkerspot the rarest butterfly in the world? Based on the criterion of numbers of individuals, no. The number of butterflies—into the thousands—that Stu's group relocates each year alone is higher than the estimated total population sizes of the very rarest butterflies. However, Bay Checkerspot populations of reasonable size have been lost, including the Jasper Ridge population, which once numbered ten thousand butterflies, and the Edgewood Park population, once numbering one hundred thousand. The Coyote Ridge population is by no means secure, but its hundreds of thousands of butterflies provide a strong foundation for conservation success.

Measured in another way, by the number of populations, the Bay Checkerspot has fallen to the lowest number possible—one! A species with a single sustainable population is vulnerable to extinction; natural or human-caused changes to a small area can wipe out the entire species. This is especially true when many threats—invasive species, climate change, nitrogen pollution, and habitat loss—all confront one butterfly species.

The threats experienced by the Bay Checkerspot's populations, its steady decline, and now its impending recovery offer special lessons about the threats faced by many rare butterflies. For now, and with the assumption that efforts are not too late, the population of Bay Checkerspots will not attain the status of the rarest butterfly. Nonetheless, given past trends, it is not unrealistic to forecast the decline of a single population to tens of thousands. When I examine the science and conservation of the Bay Checkerspot, however, I see something different. The sustained efforts on its behalf have created a scientific foundation to build a real and lasting future for the species. The investment needed to reach this point is considerable. With it, real conservation of the Bay Checkerspot can be achieved.

CHAPTER 3

FENDER'S BLUE

In 1929, University of Minnesota entomologist Ralph Macy captured a small, blue butterfly in western Oregon (Plate 4). He discovered it in the middle latitude of the state, on the western side of the Willamette valley, just west of the city of Salem. In 1931, Macy recognized that it was a new subspecies and named the butterfly Fender's Blue (*Plebejus maricopa fenderi*), after his friend and fellow butterfly collector Kenneth Fender, who had first observed the subspecies. As the relationship between this and other butterflies became clearer, it assumed its present scientific name, *Icaricia icarioides fenderi*. In the eight years following its discovery through the period of the Great Depression, the Fender's Blue was observed a number of times between Salem and Corvallis, located thirty miles to the south.

After 1937, the Fender's Blue vanished. It did not disappear for just a year, or a few years, or a decade. It was lost for the next half century and was considered extinct.

TURNING PRAIRIES INTO GRASS

The Fender's Blue was not always rare. There is no way to know its exact population size and range, but historic numbers must have been higher. It likely flew over a vast network of habitats—in a series of metapopulations—that connected the extent of the Pacific Northwest prairies. Created more than ten thousand years ago when floods followed glacial retreat, these habitats once covered one million acres in the Willamette River valley. If the abundance of the Fender's Blue matched the extent of these prairies, its numbers would have been enormous.

The butterfly's loss began in the mid-nineteenth century as a consequence of three great changes in its habitat. First, a wave of European settlement launched unabated plowing for agriculture. As is the case with all the rarest butterflies, habitat destruction is the primary cause of the Fender's Blue's decline. Two-thirds of the total area of prairie was dry and thus easily plowed and planted into crops. The other third was wet. Early settlers spared wet prairie because it was virtually inaccessible to farming, thus providing a safe haven for the Fender's Blue. However, loss of wet prairie accelerated in the 1930s, when the US Army Corps of Engineers executed water management and drainage work. By 2009 a million acres of the Willamette valley were dedicated to hay, nearly a million to wheat, and just under half a million to grasses grown for their seeds. These millions of acres of grasses are no more similar to native prairies than are cornfields or pine forests. The valley also supports many fruit, vegetable, and Christmas tree farms. The prairie now encompasses fewer than one thousand acres, scattered across the landscape in small and isolated fragments.

The arrival of settlers simultaneously launched a second stage of loss, caused by large-scale fire suppression. In conservation, we focus first on the most obvious habitat loss, which follows when people convert land wholesale from forest or prairie to cities and farms. A recurring theme among the rarest butterflies is that population decline can also occur within what appears to be the species' proper habitat. People degrade habitat by creating barriers to natural processes of disturbance, in this case fire. When fire does not occur in prairie, shrubs and trees such as Himalayan Blackberry (*Rubus armeniacus*) and Pacific Poison Oak (*Toxicodendron diversilobum*) supplant native grasses and herbs; in the Willamette valley, this includes the Fender's Blue's host plant, a species of lupine. Before settlers arrived, the Kalapuya people had promoted prairie growth through intentional burning. Large fires helped to increase bulbs of Small Camas (*Camassia quamash*) and wildlife that were important sources of food. Historians documented near-annual fires until early in the nineteenth century, when settlers largely suppressed fires to maintain the agriculture they needed to support their families and communities.

In addition to keeping out trees and shrubs, fire served as a bulwark to a third change, one that is shared with the Bay Checkerspot: the intrusion of non-native, invasive species, including Common Velvetgrass (*Holcus lanatus*), Slender False Brome (*Brachypodium sylvaticum*), Scotch Broom (*Cytisus scoparius*), and Tall Oatgrass (*Arrhenatherum elatius*), among others. These now blanket the landscape. Without fire, these plants have outcompeted native prairie grasses and herbs, including the host plant of the Fender's Blue.

My former doctoral student Erica Henry, who had previously completed her master's degree near the Willamette valley home of the Fender's Blue, described to me an irony embedded in this butterfly's story. The climate after the post–Ice Age glacial retreat apparently favored trees and shrubs. Native Americans lit fires that kept these woody plants at bay, allowing the Fender's Blue's host plant to survive. With fire suppression, we are now left with a puzzling situation: the pendulum has swung from more to less fire for an endangered butterfly that might have been extinct centuries ago if people had not intervened.

To achieve the conservation goal and reclaim fields of lupine-filled prairie in the Willamette valley, there are two paths forward. One is to reclaim areas that once supported the Fender's Blue. Purchasing land for conservation is conceivable but happens slowly, through the accumulation of small parcels; this type of conservation is a long road. The other path is to manage existing habitats properly, introducing the appropriate frequency and intensity of fire and other management. To accomplish this would require new science. However, restoration would not make sense for a butterfly considered extinct since 1937.

SCIENTISTS AND OUR BAD ASSUMPTIONS

Five decades after the butterfly's apparent extinction, the history of the Fender's Blue took a remarkable turn. In 1988, Paul Severns rediscovered the Fender's Blue. A twelve-year-old butterfly collector, Severns traveled to Coburg Ridge, Oregon, to find unusual forms of checkerspots and fritillaries. When he captured a blue butterfly, he recognized immediately that it was different in color pattern from other individuals in his collection. When he determined it to be a Fender's Blue, he did not consider it noteworthy. His field guide referenced the Fender's Blue just as it did all the other butterflies; it said nothing about its apparent extinction. Later that fall, he attended a meeting of butterfly scientists and enthusiasts who were skeptical of his rediscovery of an extinct butterfly. The next day, he brought his specimens to the meeting. He showed them to, among others, Oregon State University entomologist Paul Hammond, who found Fender's Blue populations the next spring. Where had the butterfly hidden in its lost decades?

As seems all too common among amateur naturalists and committed scientists interested in rare species (especially insects), people were looking in the wrong places. The crux of the oversight was that scientists thought they knew the Fender's Blue's host plant. They did not. It seems so simple to connect a caterpillar to its food. With rare and sometimes cryptic butterflies, however, there are too few caterpillars to easily observe feeding, and so we make assumptions based on food and habitat associations of similar species. A lesson that I have had to learn and relearn is that it is easy to draw simple but misguided conclusions about host plant relationships for rare butterflies.

The host plant of the Fender's Blue is remarkably similar to but subtly different from certain other plant species. Like many blue butterflies, the Fender's Blue feeds on lupines (genus *Lupinus*), a type of plant in the legume family, which includes peas and beans as well as many different species commonly found in nature and planted in gardens. Entire groups of lupine species look remarkably similar. Early on, Ralph Macy and others had observed the Fender's Blue around a small population of a lupine, but they had misidentified the true host plant.

In their search for the Fender's Blue, entomologists and collectors initially searched common lupines in its range, concentrating on perennial species. When the Fender's Blue was resurrected, it did not occur in association with common lupines but with a very rare species. What confounded the search was that in rare circumstances the butterfly could use a variety of host plants. These can be locally rare but have a wider global range. One suspected host plant was Spurred Lupine (*Lupinus arbustus*), found primarily in eastern and central Oregon and extending north into Washington and south into northern California. It grows one to two feet tall, and its flowers display a variety of colors from yellow to pink to blue. Another was Sickle-keeled Lupine (*Lupinus albicaulis*), found in the western part of Oregon but overlapping the range of Spurred Lupine; it grows to two feet tall and has purple flowers.

Although uncommon plants, Spurred and Sickle-keeled Lupines were not the primary food source of the Fender's Blue. That was another lupine, one that is very, very rare. The threatened Kincaid's Lupine (*Lupinus oreganus*) grows one to two and a half feet tall and has purple flowers. Although similar to Spurred Lupine, it differs in that its flowers lack long, hollow, extended spikes (or spurs). Confounding things even further was that these two species could interbreed and create hybrid varieties. The relationship between the Fender's Blue and Kincaid's Lupine was key to unraveling the story of the butterfly's biology and conservation. Here we have a rare and protected butterfly feeding on a rare and protected plant. An intriguing combination, it is one that predisposes the butterfly to small numbers. As I'll discuss later, conservation of the Fender's Blue must build on the foundation of conservation of Kincaid's Lupine.

Botanist Charles Smith identified Kincaid's Lupine around the time of the butterfly's discovery. In 1924, he described it and twenty-five other lupine species, setting them apart from the other 110 lupine species in the Pacific Northwest. Smith considered Kincaid's Lupine a variety of Oregon Lupine and named it after Trevor Kincaid, who first collected the plant in 1898. Over the past century, the designation of Kincaid's Lupine as a variety or a subspecies and its relation to other, similar species has vacillated. Some species are indistinguishably similar, including hybrid forms. This helps to explain why butterfly experts were confounded as they searched for the Fender's Blue. Without narrowing the range of lupines to this one species, the search for the Fender's Blue was quixotic.

MY FIRST ENCOUNTER

In 2011, I took advantage of knowledge about the Fender's Blue that had accumulated in the two decades since its rediscovery. I created a surefire plan to see the butterfly. I timed my cross-country trip from North Carolina to western Oregon at the ideal time of year, late spring. Washington State University professor Cheryl Schultz led me to two sites, one in wet prairie and the other dry. It was immediately apparent why these places were of low quality for agriculture and unlikely to be plowed.

Our first stop was Baskett Slough National Wildlife Refuge, located in the foothills just outside Salem, Oregon. The prairies rose out of the valley and extended up slopes in fingerlike protrusions, where the prairie interwove with the forest that dominates at higher elevations. We ascended Baskett Butte in search of the Fender's Blue and Kincaid's Lupine. The hike was not hard, a round-trip of about a mile and a half, with an elevation gain of two hundred feet. On the beautiful day of our ascension, the view over the river valley from the crest of the butte was stunning. The landscape was open prairie with rocky terrain, and grass and hay fields blanketed the valley floor below.

More foreboding was the tall, dense growth of poison oak. Before our trip, Cheryl had warned us repeatedly not to step off the path in the wrong direction, as the wrath of these plants awaited. Just beyond this peril and in tantalizing view, Kincaid's

Lupine was blooming with vigorous purple flowers and was thick with leaves.

I had come to Baskett Butte with a team of biologists that included professors, graduate students, and research technicians. We searched for Fender's Blues, all of us full of anticipation. For those who had never seen the Fender's Blue before, this was a chance to see a unique and endangered animal. Our group was full of inquiry, talking about the science of butterflies and plants as we searched the lupines and admired their flowers.

Although I had planned the trip impeccably, a cool and rainy spring and early summer foiled our efforts. The poor weather conditions delayed the Fender's Blue's development, and I failed to see a single adult butterfly. My disappointment lingered for years.

In the context of my hunt for the rare butterfly, however, my efforts were not completely in vain. If there were not adults, there must be caterpillars (hard to find) or chrysalides (almost impossible to find). My entire group knelt and hunched over lupine plants and undertook the tedious search for caterpillars by flipping leaves. Caterpillars find refuge on leaf undersides during the day, where they rest before feeding again. There were many lupine leaves to flip and very few caterpillars to discover.

The caterpillars are small, green, pudgy, and soft (Plate 5, top). Like the larvae of many other butterflies, they conceal themselves from predators by adopting the color of their host plant. They have no discernable streaks, markings, hairs, or spikes. At times, they will crawl out from leaves onto flower stalks, where they sit in contrast with the petals. People will often observe them together with their partners, ants that feed on sugars the larvae secrete while at the same time protecting the caterpillars from potential predators.

We spent a couple of hours searching. I admit that I was hopeless in my search for caterpillars. My attention was distracted by any small shadow that passed over, as I hoped I'd see an adult. My eyesight is not as good as it once was, and I am not skilled at detecting small, inanimate animals hugging up against plant leaves. We each searched dozens of plants and hundreds of leaves. Finally, one of my collaborators yelled out to the rest of us. We gathered around to see one good-size caterpillar (Plate 5, bottom). We managed to see a few more that day. I'll admit to some excitement as I observed a very rare species at a difficult-to-find life stage. Yet, as I had seen no adult, the trip fell far short of my expectations.

Although I was in search of an adult butterfly, the underlying reason for our visit to Baskett Butte was for a broader research agenda. Cheryl and I participated in a project to study the effects of restoration on butterfly population sizes. We saw most of what we'd come to see: dry prairie, Kincaid's Lupine, Fender's Blue caterpillars, and some habitat management. We also observed techniques Cheryl's lab used to study butterfly behavior, which I will discuss in more detail below.

Another goal that day was to see the variety of landscapes in which the Fender's Blue lives. After a short lunch, we traveled sixty-eight miles south, essentially traversing the butterfly's north–south range, to Willow Creek Preserve in Eugene, Oregon. The five-hundred-acre site, preserving wet prairie, is run by the Nature Conservancy. The area is large and relatively flat, with butterflies and their host plants appearing in areas of the preserve that rise a bit out of the wetter terrain. Despite my lingering hope of seeing at least one adult butterfly, on this day we saw no Fender's Blues flying at Willow Creek. What we did see close-up were examples of how and where efforts were underway to restore prairie.

RESTORATION BEGINS WITH PLANTS

One point that Cheryl impressed on us at each stop on our visit was that any effort to recover the Fender's Blue had to be built on successful recovery of Kincaid's Lupine. While some rare butterflies depend on host plants with relatively wide distributions, the Fender's Blue's host plant is threatened. Even so, Kincaid's Lupine has a broader distribution than the Fender's Blue. Its range encompasses more than 150 populations, five times the number of Fender's Blue populations. There are two reasons the Fender's Blue does not occupy every site with Kincaid's Lupine. First, a Kincaid's Lupine population can be very small, comprising clusters of just a few plants along roadsides. Such populations are unable to support a robust Fender's Blue population. Second, Kincaid's Lupine grows outside the Fender's Blue's historical range. Nearly all of the plant populations occur in the Willamette River valley of western Oregon, running the entire north–south extent from Portland to Eugene. The conservation goal for Kincaid's Lupine recovery is to maintain suitable conditions to support populations and to conserve new habitats to expand its range. The conservation goal for the Fender's Blue is to increase populations to corresponding areas.

Two problems stand in the way of Kincaid's Lupine range expansion. The first and easier problem to solve is to acquire the botanical knowledge needed for restoration success in new areas. Nursery professionals have already learned to propagate Kincaid's Lupine for planting in new locations. The second and more challenging problem is land acquisition for restoration. Kincaid's Lupine occurs within restoration areas that total six hundred acres. Within them, Kincaid's Lupine foliar cover, summed up across all areas, amounts to an area of just 3.8 acres.

The plant will need to occupy larger areas to sustain its and Fender Blue's recovery.

Kincaid's Lupine can colonize new areas as plants or seeds. Cheryl conducted experiments to test restoration methods by sowing seeds into prairies. She laid the groundwork by killing invasive plants. She found that one form of site preparation was most effective: overlaying the soil with plastic sheets and thus creating a greenhouse that harnessed the sun's rays and heated the soil, a technique called solarization. This performed better than tilling soil or removing key nutrients. Even so, less than 10 percent of Kincaid's Lupine seeds germinated, and just a few plants set root and grew for the duration of the study. This research led Cheryl to conclude that restoration of Kincaid's Lupine would have higher success through the transplanting of seedlings, rather than the sowing of seeds.

Once Kincaid's Lupine is established, restoration success depends on proper growing conditions. Conservation managers adopted three strategies to knock back the shrubs, trees, and invasive grasses that choke out native plants such as Kincaid's Lupine in the Oregon prairie. First, they restored fire. Fires lit in late summer to early fall were ideal for Kincaid's Lupine, causing plants the least harm. At this time of year, the plant's leaves turn brown, and its live parts are below the soil. A challenge of fire is that strong winds might carry flames in unwanted directions, putting homes and cities at risk. Fire now occurs in the habitat of many but not all populations of the endangered plant. The second way the managers reduced the density of unwanted plants was to replace fire with mowing. Although mowing could help accomplish restoration goals, it can actually harm Kincaid's Lupine by reducing growth of flowers and leaves. Despite this possible risk, nearly all habitats with Fender's Blue populations are mowed every year. Third, the conservation managers adopted a brute-force approach and physically weeded out unwanted invasive plant species. This task required great effort by many people. Despite the high level of physical labor, there is now an aggressive campaign to eliminate invasive plants. Kincaid's Lupine will require some combination of these strategies, considered together, to persist. Ideally, the strategies will align with the restoration needs of the Fender's Blue.

THE KEY INGREDIENT: BUTTERFLY NATURAL HISTORY

In historic or restored areas that harbor Kincaid's Lupine, a significant challenge is to expand Fender's Blue populations to these areas. As of 2018, there were about three dozen known populations of the butterfly. They occupied a range extending eighty miles from Eugene in the south to just north of Salem. Expansion of the range could occur in one of two ways. People could take a page from Bay Checkerspot restoration, supplant natural dispersal, and move eggs, caterpillars, or adults to locations that have Kincaid's Lupine but no butterflies. For the Fender's Blue, this strategy is considerably more difficult. The Bay Checkerspot has two advantages over the Fender's Blue. It occurs within one large population, and it lives in clusters during the egg and caterpillar stages of its life. Beyond the challenge of finding individual Fender's Blue eggs and caterpillars to move, it is harder to justify taking too many individuals from its much smaller populations. Alternatively, conservationists could design a network of planting areas such that each population of Kincaid's Lupine is within the range of the Fender's Blue's dispersal capability. Overcoming these barriers requires new science and innovative management.

Restoration of the Fender's Blue is contingent on accommodating the butterfly's biology. This brings me back to my visit to Baskett Slough in 2011 and to the true purpose of the trip. I was interested in seeing Cheryl's research on the biology and restoration of the Fender's Blue. Cheryl has served as the species' primary scientist and steward and has invested in the research and conservation needed to understand how to protect and recover its populations. One lesson I've learned from her is how little we understand about the natural history of rare butterflies and the significant dedication needed to acquire that knowledge. By visiting sites with Cheryl, I knew I could learn about the restoration and recovery of the Fender's Blue as well as more general lessons about the conservation of endangered butterflies.

Cheryl began her efforts on behalf of the Fender's Blue for her PhD dissertation, "Ecology and Conservation of Fender's Blue Butterfly." This tome ran the gamut, conveying Cheryl's understanding of everything from the Fender's Blue's behavior in and near small, fragmented prairies, to its dependence on plants for caterpillars and adults, to restoration with fire. Twenty-five years have elapsed since she began this research, and Cheryl is still focusing her career on this butterfly. Thanks to her efforts, and because she has stayed with it for the long haul, there are signs of significant progress toward the Fender's Blue's conservation.

Until 2011, Cheryl had found that individual populations of Fender's Blue varied in size from as high as two thousand individuals to as low as just a few butterflies. The total number of butterflies across all sites fell within the range of two thousand to six thousand. Cheryl initiated efforts to accumulate the science needed to promote recovery.

In 2003, Cheryl teamed with Paul Hammond to predict the future sizes of every Fender's Blue population. They knew that Fender's Blue population sizes, as do those of butterflies in general, fluctuated in large swings, with very high levels in one year followed by very low levels the next. They used this variability to visualize population sizes into the future. Even in the best habitats, fluctuations in rainfall and temperature influence how many eggs each Fender's Blue lays and how many caterpillars become adult butterflies each year. At the extreme, downward fluctuations can be so large that they result in extirpation.

Cheryl and Paul found that in the absence of restoration, only one population of Fender's Blue had a 90 percent chance of surviving this century. Other populations had a fifty-fifty chance, or less, of survival. Natural variation compounds the threats of loss of habitat, invasion by exotic species, and termination of fire, predestining populations to be lost.

In her years working with land managers to identify and implement methods to conserve the Fender's Blue, Cheryl has found it difficult to translate models that predict large fluctuations in nearly all butterfly populations. The concept is too abstract. Given that the stakes for the butterfly are so high, as an alternative, she established a minimum size of Fender's Blue population that is required for the butterfly's populations to recover. Informed by a variety of her studies and her frequent interactions with those tasked with on-the-ground conservation, Cheryl set the lowest threshold for any one population at one thousand butterflies.

Two decades of research have allowed Cheryl to identify how to integrate the Fender's Blue butterfly's biology to guide restoration. She studies butterfly behavior to fine-tune restoration dedicated to areas within existing sites and to restoration of new sites; population demography to focus restoration on the types and locations of plants, including nectar and host plants; and greenhouse-raised caterpillars that can become the seeds of future populations.

On the day I visited Baskett Slough in 2011, our focus was on the Fender's Blue's behavior. While traversing the preserve, we eventually emerged through a forest gap to arrive at a large open grassland. We looked across an area that was relatively flat but rose with a shallow slope. It was here that we first observed the methods Cheryl used to examine butterfly populations and behaviors. Cheryl studies the Fender's Blue in flight to understand and predict how it is able to move from plant to plant within a prairie or from prairie to prairie within a landscape. These studies are important in helping us understand how to restore habitat that a butterfly moves within and how to restore neighboring habitats that are in the target range of the butterfly's dispersal. In the context of the wider landscape, Cheryl's group has found that forest and other habitats that do not suit the Fender's Blue or Kincaid's Lupine impede butterfly flight.

As the Fender's Blue was not flying the day of our visit, we tested methods used in studies of the Fender's Blue on a related look-alike, the Silvery Blue (*Glaucopsyche lygdamus*). Cheryl has employed one of these methods for more than two decades, over which time she has followed the paths of thousands of butterflies for short distances until they stop, turn, or perch. Each of these distances is considered one step along a path, and a path comprises ten to twenty steps. When students track a butterfly, they place a flag attached to a thin wire to mark the end of each step. After the students complete marking a path, they record the distance and direction of each step. A key piece of equipment, used to record and organize the data, was a Tandy Pocket Computer, manufactured in the 1980s. I chuckled when I saw this vintage technology, but I realized quickly that Cheryl had adopted it for several of its features, including its small size and the ease with which students carried it. The Tandy allowed them to record data in a particular way: they could press a button each time a butterfly turned, stopped, or fed, and that motion put a time stamp on each step. A student could watch a single, moving butterfly and easily record data. Cheryl's former master's student and my former PhD student Erica Henry told me the story about how each winter a lab member would sort through nonfunctional Tandy Pocket Computer units and send them to someone in California who specialized in repairing these devices. This repairman has now passed away, and the lab's use of Tandy Pocket Computers may follow.

When our group tried the path-tracking method, we released Silvery Blue butterflies at one end of the field and followed them, dropping flags as they flew. The process of acquiring the data could be frustrating, as the butterflies took a long time to fly short distances measured a few yards at a time.

Cheryl, her students, and her collaborators used the data they collected to ascertain where Fender's Blues go within and among areas that harbor their host lupine. They found that Fender's Blue butterflies spent most of their time in single stands of lupine that measure yards across. When the butterflies decided to leave, they typically moved slowly and remained close to their originating site, within about ten yards. This defined the ideal range of restoration within a habitat area.

When Fender's Blues entered unfamiliar (and potentially dangerous) terrain, they shifted into a higher gear. The Fender's Blue can disperse distances measured in kilometers. They biased their movement away from dense forest and open woodlands without lupine and toward open woodlands and prairies with lupine. Yet dense woods did not form a perfect barrier. Cheryl's group ascertained that land managers should locate areas harboring dense lupine no more than two kilometers (1.2 miles) apart. These were the types of data needed to map out a plan for butterfly recovery; they determined the distances between new parcels for acquisition and restoration. These would be in the target range of the Fender's Blue's capacity for dispersal.

A different study of Fender's Blue behavior exposed a counterintuitive scenario in which butterflies might be attracted to restoration sites where their populations would actually fare worse. Cheryl's former graduate student Leslie Rossmell investigated restoration in sites immediately adjacent to an existing butterfly population at the other site I visited in 2011, Willow Creek. Mowing and weeding to reduce unwanted species led to increases in Kincaid's Lupine flowers, which attracted adult butterflies. However, the Kincaid's Lupine that was seeded in the area did not achieve the density of leaves, essential to caterpillar growth and survival, found in the best sites. The new flowers attracted Fender's Blues away from neighboring habitat with high-quality patches of lupine, and they proceeded to lay large numbers of eggs there. It was jarring to watch populations in the original, unrestored habitat actually decline following restoration.

Any hopes that the Fender's Blue will establish in distant populations of Kincaid's Lupine will require butterflies to be moved there. One approach could be to raise butterflies in greenhouses, an approach that Cheryl has found to be difficult. There is no straightforward way to mimic natural environmental conditions in greenhouses. This is especially true when climates vary through summer and winter during the butterfly's life cycle. Cheryl and her group tested greenhouse methods with a blue butterfly related to the Fender's Blue. What she found gave her pause. Captive-reared butterflies were smaller than those raised in the wild. Smaller butterflies performed more poorly. Only about 10 percent of greenhouse-raised butterflies survived from egg to adult. Still, this rate was higher than rates expected in nature, as greenhouse settings support limitless food and exclude predators. A compromise strategy could be to restore prairie and then reintroduce caterpillars to grow and mature as they would in wild settings, as was done for the Bay Checkerspot. Given the limitations of captive rearing, Cheryl's group has turned to the possibility of moving caterpillars directly from current habitat to restored sites.

BURNING BUTTERFLIES?

The bestowal of the scientific name *Icaricia icarioides* to the Fender's Blue and related subspecies has always struck me as prescient. Some butterflies are named to convey a feature of their morphology or environment. This was not so for these blue butterflies. In the case of *Icaricia icarioides fenderi*, the metaphor of the Greek mythological character Icarus serves as a model for the subspecies' struggles with fire.

Cheryl's, and others', most important realization for Fender's Blue restoration, one held in common with the restoration of other rare butterflies, was that we need to overcome our instincts to prohibit ecosystem disturbance. In my own research, I have fallen into the trap of trying to invent restoration that was "gentle." I have found, time and again, that I am not alone.

There is one imminent need for Fender's Blue restoration: fire. In this arena, scientific advances for Fender's Blue conservation have served as a model for the management of a number of other rare butterfly species. As with all cases of restoring disturbance, the issue of returning fire to an ecosystem is controversial. Pacific Northwest prairies depend on fire to keep out trees, shrubs, and invasive species. Without fire, the habitat is lost. Simultaneously, fire kills caterpillars crawling in its path. Application of fire balances on a knife edge. Either too much or too little fire will lead to the extinction of the Fender's Blue.

The Fender's Blue benefits from tough love. Research at Baskett Slough has shown that the benefits of fire outweigh its potential costs. Cheryl's group conducted rigorous experiments in which they burned some areas and left other, control sites unburned. In comparisons between these two treatments, the contrast was clear. Fire improved habitat, stimulating Kincaid's Lupine to flourish. This benefited the butterfly, and female Fender's Blues laid on average five times as many eggs in recently burned areas. The capacity of these populations to grow was inherent, as each adult female had the ability to lay around 350 eggs. Caterpillars fed on plants and overwintered in the leaf litter near or at the base of their host plant.

It is not too difficult to predict the negative effects of fire. The Fender's Blue is nearly twenty times more likely to survive on unburned host plants. Eventually, this loss is made up for by the greater abundance of host plants after a fire. Tufts University biologist Norah Warchola discovered another benefit. She found in burning experiments that caterpillars recruited their important ant mutualists in recently burned sites. After burning, Fender's Blue populations often declined initially but then grew at faster rates in following years. In this scenario, "just right" meant killing some individuals within a population in the short term (caterpillars living in sites that are burned) for the benefit of others in the longer term (more eggs and lupine regrowth after fire).

To reconcile these costs and benefits, Cheryl teamed up with Tufts University professor Elizabeth Crone. The two are longtime collaborators in Fender's Blue research and conservation studies. They created models of butterfly population growth, basing the models on actual measures of butterfly survival and reproduction. They then used their models to test different fire management scenarios. They varied two aspects of the environment: the fraction of the landscape burned (from one-eighth to one-half) and the time between fires (from one to five years). Evaluating all possible combinations, Cheryl and Elizabeth arrived at two conclusions. First, the amount of habitat burned could be too small; burning more than a quarter of the landscape was best. Second, the frequency of fire could be too low, especially if time intervals were greater than two years. Although many of the strategies the two scientists considered in combination would work to conserve the butterfly, the model pegged the ideal rate of burning at one-third of the landscape every year. Although incinerating large parts of a Fender's Blue population is undesirable, restoration with fire promotes future growth.

One thing that has impressed me about Cheryl's efforts is her ability to transfer basic science to on-the-ground restoration. I asked Cheryl: Have land managers implemented the study's recommendation for burning? She told me that they have in some but not all sites. A key point she conveys to land managers is that they need to burn a third of the *high-quality habitat*. This is not the same as burning a third of a field. Cheryl has tried to impart her knowledge of how a butterfly sees high-quality habitat. Kincaid's Lupine is not everywhere in a grassland opening; high-quality habitat is the much smaller subsection of a grassland that contains lupine.

Because of proper management, the total population size of the Fender's Blue rose from about three thousand butterflies in 2000 to about twenty-nine thousand butterflies in 2016. There was lots of year-to-year variability in numbers, including a drop back to thirteen thousand butterflies in 2017. Scientists expect large fluctuations in butterfly populations caused by variability in environments and uneven effects of land management. The higher numbers corresponded to increased efforts in habitat restoration, including a combination of fire, mowing, removal of invasive species, and planting of nectar plants. Ongoing efforts are refining the optimal combination of these interventions. Still, Fender's Blue population sizes are headed in the right direction.

As I've studied Cheryl's research on disturbance management for the Fender's Blue, I've learned that her management recommendations can be applied more generally to the conservation of many rare butterflies. For example, another rare butterfly, the Bartram's Scrub-Hairstreak (*Strymon acis bartrami*), appears to require fire. It is a small silvery butterfly found in the Everglades and Florida Keys. In 2014, the US Fish and Wildlife Service added it to the list of threatened and endangered species. In the small areas of its habitat that remain, the butterfly depends on fire for the recovery of its caterpillar's host plant, a small shrub called Pineland Croton (*Croton linearis*) that is easily overgrown by larger shrubs and trees. In a vacation destination like the Keys, fire is prohibited due to concerns of its escape near homes. On Big Pine Key, the best habitats I've observed for the Bartram's Scrub-Hairstreak are not the places where the habitat has been actively managed. Rather, they are former campsites where people had accidentally set fires. Fire management for this species is in its incipient phases; perhaps its managers could learn a lesson from the efforts made on behalf of the Fender's Blue.

With limited resources for burning, mowing, and planting, the people involved in Fender's Blue conservation must optimize efforts. Cheryl's group created models of populations in computer-generated landscapes to identify the best configurations of restored sites. These results will help to guide where new parcels should be purchased to complement habitats that already exist. Cheryl's simulations included areas that were either large or small and that were either connected or distantly isolated from each other. Her group found that larger areas performed well whether they were located near to or distant from existing habitats. On the other hand, small areas were doomed unless they were connected to other parts of the population. A key lesson was that the importance of dispersal between sites depended on the performance of a population within a given site. For a distant population between which exchange via dispersal is impossible, a minimum conservation area likely exceeds fifteen acres.

SCIENCE AND STAKEHOLDERS

Because of research by Cheryl and her collaborators, and because of interactions between Cheryl and stakeholders in land management and in state and federal agencies, the total number of Fender's Blues can, in a good year, reach twenty-nine thousand butterflies. Investment in science and restoration for the Fender's Blue has paid dividends. Although this butterfly is still rare, its recovery has shifted it out of the running for the title of the rarest. Effects of fire in restored areas last for three years. At that rate, the cycle of fire management will have to continue at a steady clip. Cheryl's research provides an example of the sustained research effort needed to understand the rarest butterflies at a level that will inform their recovery.

Fender's Blue science and management have tipped the balance toward the butterfly's resilience. Looking back over its entire known history of eight decades, it has mainly declined. The Fender's Blue was rare when discovered, as its habitat had already experienced loss of over 99 percent. Nearly immediately after its discovery, it seemed to go extinct. Once it was rediscovered, ongoing development in combination with habitat degradation caused by removal of essential disturbances reduced the Fender's Blue's habitat and population size even further.

The tides now appear to be turning. Powered forward by dedicated research, committed protection, and active conservation, population sizes are rising. This is the result of the coordinated efforts of scientists and land managers dedicated to conservation. For most of the time while writing this chapter and book, I remained uncertain as to whether the Fender's Blue was the world's rarest butterfly. At the very least, it appeared a contender. Carefully planned restoration in a variety of forms has enabled the butterfly to bounce back rapidly. This work shows the potential for restoring threatened butterflies and other animals and plants, and reversing the course on which global changes previously set them. If there is one lesson I draw from the Fender's Blue, it is that preserving the rarest species can happen with dedicated effort in research and management. It inspires me with optimism that conservation is possible for other butterflies.

The recent success of Fender's Blue conservation raises new questions. Can it persist without ongoing management intervention by people? For now, the Fender's Blue is conservation-reliant. Because its habitats depend on fire, and because fire is not going to reappear in its habitats except in the most controlled settings, I cannot see this changing. A great aspiration is to achieve self-sustaining populations, but more research and conservation are needed to get there.

My attempt to see an adult Fender's Blue butterfly in 2011 failed, but I clung to the possibility of seeing it. I made a return visit to Oregon in 2016. This visit was the opposite of the previous one, when a cool and damp spring had delayed butterfly flight. This time, climate change in concert with a strong El Niño current had produced the warmest year on record. As did other butterflies across North America, the Fender's Blue responded by emerging nearly two weeks earlier than usual. The timing of my trip was impeccable, as I'd positioned my visit at the peak of Fender's Blue abundances. I visited Baskett Butte on a warm and sunny day, the perfect conditions for butterflies to fly. Within minutes of arrival, I observed my first Fender's Blue. I spent an hour walking through the site, and I observed at least a dozen Fender's Blues. Although I had already realized my dream of seeing the Fender's Blue, more surprises were in store. As I photographed a Fender's Blue that had alighted on a Kincaid's Lupine, my collaborator observed another land on my back. Another endangered butterfly had made it onto my life list.

CHAPTER 4

CRYSTAL SKIPPER

In 1978, Eric Quinter, a specialist in butterflies and moths at the American Museum of Natural History in New York City, collected insects on barrier islands along the North Carolina coast. His focus was moths. As he explored the dunes near Atlantic Beach, he netted a few butterflies and added them to his collection. These belonged to a group called skippers. Because of skippers' small size and brown to rust-orange color, most people who see them in their gardens think they are moths. Clues that the Crystal Skipper is not a moth include its flight during the day, its landing position with wings folded behind the body, and the definitive characteristic of butterflies: bulb-shaped (not feathery, as in moths) antenna tips. Quinter did not suspect that the butterflies he netted that day were unusual.

Five years later, Quinter showed his collection to Smithsonian entomologist John Burns, a specialist in the skipper group of butterflies. He immediately recognized the Atlantic Beach specimens as something different. Were they a different form of a more common, mainland butterfly species? Were they a new subspecies? Were they different enough to be classified as a new species? Finding the answers to these questions was especially difficult because of uncertainty in classification of the two closest species for comparison, the Dusted Skipper (*Atrytonopsis hianna*) and the Loammi Skipper (*Atrytonopsis loammi*). Quinter's new butterfly population on Atlantic Beach had thrown an additional wrench into the uncertain classification of these butterflies, and its status confounded scientists for decades.

With Quinter's butterfly locked in scientific purgatory, scientists and the US Fish and Wildlife Service gave it an enigmatic, temporary name: *Atrytonopsis* new species 1. The conferment of the name recognized that the new population from Atlantic Beach differed significantly from its closest relatives. It differed in four ways. First, it was geographically separate. The Loammi Skipper had a narrow distribution in a few places south of North Carolina, mainly in Florida. Although the Dusted Skipper had a much broader distribution across the eastern United States, including on the mainland of North Carolina, it was separated from the new population by a three-mile-wide estuary. Second, its caterpillars fed on a different host plant. The new population did not feed on Little Bluestem (*Schizachyrium scoparium*), as the Dusted Skipper did. A common grass endemic to large areas of prairie and fallow fields, Little Bluestem is not frequently found on coastal dunes. The caterpillars of the new butterfly fed on a different species, Seaside Little Bluestem (*Schizachyrium littorale*). Third, unlike the other two butterflies, which produce one generation per year, the new butterfly produced two. Fourth, the new butterfly looked different. All three types of butterflies were brown. Set against the dull brown background, each species had a series of crystalline spots. In the Dusted Skipper, the most common form, the hindwings had one small white spot and frosty margins. The hindwings of the Loammi Skipper had a series of white dots, a few near the body and an arc of spots farther away. In contrast, the new butterfly had a series of large spots on the hindwings and on the forewings (Plate 6).

Did these observations provide evidence compelling enough to justify classifying the Atlantic Beach population as a new species? At a meeting in 2000, Burns made the most concrete statement to that date suggesting that they did. Still, this fell short of formal recognition in a scientific paper. With momentum seeming to accumulate in favor of an official species name over more than three decades, I found it hard to understand how the process could last so long. As the process played out, I relegated myself to the role of a passive but interested observer. I felt some sense of urgency, because the butterfly's range was so small and its abundance was so low that as a new species or subspecies it would be highly threatened and in need of protection.

The butterfly's enigmatic status seemed to preclude its inclusion in my search for the rarest butterfly. I had assumed that my search had fixed boundaries. Intuitively, "rarest butterfly" referred to a single species or subspecies that had a small number of individuals or occupied a very limited area. The rarest butterflies should be able to appear in a butterfly field guide with their own photo and description. Perhaps I should have come to realize earlier that the identity of a species is not fixed. By any standard, *Atrytonopsis* new species 1 was a very rare butterfly. I then had an epiphany: I could turn the uncertainty of the population's status into an opportunity. If it was, in fact, a new species or a new subspecies, I could consider it in my assessment of the rarest butterflies.

My former graduate student Allison Leidner recognized this opportunity. *Atrytonopsis* new species 1 was rare enough that she decided to make it the subject of her PhD dissertation research. Allison began her studies while the butterfly's status was still in flux. Resolution would not happen until after her dissertation was complete. Yet, Allison had the foresight to realize that *Atrytonopsis* new species 1 would eventually become a species or subspecies.

During her dissertation research, Allison helped to guide the butterfly's fate when she gave it a common name. Unlike a formal scientific name, which is generated by a taxonomist and then vetted in peer review, common names arise informally. Allison was frustrated when trying to explain the conservation needs of a not-yet-named butterfly to agencies in charge of land protection or to interested beach-goers. She decided to affiliate the name with the region where it lives, a thirty-mile stretch of dunes immediately adjacent to the Crystal Coast. Allison settled on "Crystal Skipper." In addition to the association with the location, the name described the distinguishing features of its otherwise brown wings, a series of small crystal-like markings. As discussions about whether to describe the Crystal Skipper as a new species dragged on, at least the common name became entrenched.

BEACH TRIP

While issues of taxonomy lingered, Allison made important scientific advances that she used to create the foundation for conservation. As a beginning PhD student in the spring of 2005, Allison drove me to the coast to find the Crystal Skipper; it would be my first sighting. It was not hard to motivate me. I jumped at the opportunity to see a rare butterfly and to examine threats to it. We drove from my home in Raleigh, North Carolina, three hours southeast to Morehead City, where we crossed Bogue Sound via the mile-long Atlantic Beach Bridge. The nearest Crystal Skippers had been discovered on the eastern end of the thirty-mile-long barrier island to which the bridge led, past the town of Atlantic Beach and within four-hundred-acre Fort Macon State Park, site of a US Civil War battle in 1862.

As I scanned the beachfront setting, part of me wished I could follow the thousands of tourists to the same destination—the beach. The southern border of the entire park is a mile-long, hundred-yard-wide beach with white sand spread evenly across it and a gentle slope into the water. The beach was tantalizingly close.

When I turned around, I saw Crystal Skipper habitat behind the dunes. Crystal Skipper caterpillars eat grass that grows just over the edge of the first dune back from the beach (Plate 7). The suitable area is a swath of land that is 50 to 150 yards wide, with the beach on one side and shrubs and maritime forest on the other. Even in ideal circumstances, the protected area may be fairly large, but the habitat available to the butterfly is tiny.

Instead of a day at the beach, I launched into the most excruciating research I have done in my career as a butterfly biologist. To find the Crystal Skipper, I walked along dunes that stop at the beach's edge. Standing at the top, I looked down on the beach, ocean waves, and people. The dunes appeared innocuous. They were not. Temperatures were doubly intense—even hotter than one would expect with an air temperature of 95 degrees Fahrenheit. The atmospheric temperature was typical of what I was acclimated to in the southeastern United States. But in the dunes, the sun's rays beat down from above *and* reflected even more heat up from the sand below, creating an oven effect. From above and from below, the sun and sand were baking hot.

Another aspect of the physical environment made the effort to find Crystal Skippers difficult. When I tell people that I study butterflies, they immediately envisage someone running across an open field with a net. These oceanside dunes did not create an environment hospitable to such frivolity. They were grassy to some extent, and they appeared idyllic. The terrain was deceptive, however, and the work was demanding. The dunes were no more than thirty feet in elevation. Nevertheless, to observe or catch fast-flying, enigmatic butterflies required that I run up and down the dunes at a rapid, steady clip. The slopes provided unstable footing on sinking sand that intensified my toil. There were yet other hazards. Not infrequently, I felt a sharp, stabbing pain as something pierced my calves. Unbeknownst to me, as I ran, my boots were catching on and ripping entire lobes from Dune Prickly Pear (*Opuntia pusilla*) plants. When my foot advanced forward, the long spines impaled the back of my calf. As my focus was on butterflies, this happened countless times before I realized that the plants were the culprit, and their spines were skewering me.

My efforts to study the Crystal Skipper solidified a competitive advantage: I'm willing to work in places most people avoid. I have an unusual capacity to tolerate harsh environments. I have relearned this about myself at each step in my search for rare butterflies. When opportunities arise to advance science and conservation, I will be there. Whatever the conditions, I am easily drawn toward opportunities to study rare butterflies. My interest in rare things is near limitless, and I jumped enthusiastically into my first search for the Crystal Skipper.

My first sighting of a Crystal Skipper was a blur—literally. Allison enthusiastically pointed in the direction of a flying Crystal Skipper. I saw nothing. I focused more intently in the direction Allison looked with her keen eyes. She pointed at another one. I observed a brown insect pop up from the vegetation and fly like a bolt above the two-foot-high grasses and over the top of a dune. I needed a new approach. These small brown butterflies sat on grass blades or near the sand. When disturbed, they flew quickly to a new location. I waited for Allison to find a perched Crystal Skipper and was finally able to sneak over to see an unremarkable butterfly. Even so, my first sighting was a great personal achievement.

The location of my first observation of the Crystal Skipper, Fort Macon State Park, marked the eastern bookend of its range. From the western end of the park, dense beachfront development chopped up the habitat running southwestward to the other bookend, Bear Island, part of Hammocks Beach State Park, located across Bogue Inlet from Emerald Isle. At 892 acres, Bear Island is twice the size of Fort Macon. It is remote and uninhabited, accessible only by crossing the inlet in a boat. The sole infrastructure on the island consists of a bathhouse, a snack stand, ranger quarters, and maintenance buildings. Overnight stays require a tent and a sleeping pad. These features combine to make Bear Island the most beautiful barrier island off the coast of North Carolina and perhaps the entire East Coast. It is, for now, a secure and largely pristine habitat for the Crystal Skipper.

I had difficulty finding Crystal Skippers—even in these parks—but my children had more luck. I conspired to take my family to Bear Island with Allison and a team of undergraduates. While I was out studying butterflies, my family enjoyed the beach. To get to the beach, they took a short path down a sandy trail that crossed the island's width. My travels took me on a perpendicular path along the length of the beach, several miles up and down dunes. At times, our paths crossed. One time when we met, my six-year-old daughter, Helen, exclaimed, "Daddy, I caught a butterfly." I smiled, pretending to play along, but Helen insisted. My three-year-old son, Owen, gestured, pointing at Helen's swimsuit. A Crystal Skipper, its proboscis extended as if to lap up nectar, was attracted to her bright, floral-print swimsuit. To Owen and Helen, the Crystal Skipper did not seem so rare.

THE BOOKENDS HOLD FEW BOOKS

In contrast to the two parks, the rest of the butterfly's thirty-mile range is heavily degraded. Habitat has been lost in three ways. The first was by dune stabilization. Left untouched, dunes move naturally. Sand accretes on one end of a barrier island just as it erodes from the other. Dune creation is a key source of habitat creation, as Seaside Little Bluestem and other native plants colonize the new areas. Dune loss takes away older areas that have become dominated by shrubs and trees that are unsuitable for the butterfly. This dynamic landscape is beneficial to the Crystal Skipper and other wildlife. However, shifting dunes are inconvenient to homes and roads. In addition, as the positions of dunes change, so do the positions of inlets. This affects bridge crossings and channels. People invest heavily in stabilizing dune locations. As a consequence, the dunes that remain diminish in quality for the Crystal Skipper over time. Once in place, infrastructure for (and history of) dune stabilization is difficult, if not impossible, to reverse.

Two other causes of habitat degradation, development and introduction of invasive plants, are easier targets for habitat conservation. Dense beachfront properties blanket the area between the bookends, making development the single greatest cause of the Crystal Skipper's decline. There is one notable exception in this area, half-mile-long Salter Path Park, and even it has suboptimal habitat. During my surveys through the grassy dunes at Salter Path, I occasionally emerged onto sandy footpaths through the protected area that connected homes to the beach. The trails formed a dense network, about one trail for each house that bordered the park. Lining the edges of these trails were tangled thickets of invasive plants, which thrived in heavily disturbed areas and overtopped Seaside Little Bluestem and other native plants.

The Crystal Skipper population's persistence and growth will depend on restoration in the areas between the large parks. When I first visited these barrier islands, I wrote off the areas outside the parks. I saw almost no space between beach homes. Then Allison took me to empty lots between and small, undeveloped areas in front of vacation homes. To my surprise, she showed me Crystal Skippers there. These places could still support the host plant and small populations of the Crystal Skipper.

Allison and I wondered: Could an initiative to build beachfront property back from the dunes create a new norm? Such a development plan would favor butterfly populations, even as it runs counter to what beach-goers find most appealing. From a beachfront home on top of the dunes, vacationers can virtually roll from their deck onto the beach. This kind of development stamps out populations of the Crystal Skipper.

It was on that trip that I had an epiphany: people and butterflies could coexist. Siting homes back from the dunes reduces their exposure to hurricanes. Hurricane season on the southern Atlantic coast lasts from June to October. Houses positioned above the beach, with no dune or vegetative barrier, are the most vulnerable to hurricane damage. Compared to areas farther south on the Atlantic and Gulf coasts, where massive storms like Hurricanes Andrew and Katrina hit, North Carolina's storms appear tame. Yet, in my near two decades living in North Carolina, I experienced powerful hurricanes, including Floyd, Matthew, Florence, and others. When they make landfall, hurricanes can wash exposed houses away. What is in the best interest of the butterflies, maintaining grassy dunes, is also in the best interest of beach-going people. Allison and others have grappled with how to communicate this to landowners and city planners. People could conserve butterflies as a spillover effect of siting houses in their own best interest.

People harm butterflies in other ways. They bring with them familiar plants that transform yards into something tidy and in some ways attractive, akin to their suburban neighborhoods. They plant species that are not native to barrier islands, to North Carolina, or to the southeastern United States. Kentucky Bluegrass (*Poa pratensis*) creates the facade of a lawn, and woody vines such as Beach Vitex (*Vitex rotundifolia*) are ornamentals planted to stabilize dunes. These are aggressive, overgrowing native plants, and their presence results in the exclusion of Seaside Little Bluestem and the Crystal Skipper.

A new approach to landscaping could reverse the influx of invasive species and restore native plants. I have observed the Crystal Skipper in small plots of native vegetation around homes. My eyes at once lit up when Allison showed me Crystal Skippers in someone's backyard, off the sand dune and three rows of houses back from the beach. The owners had maintained native vegetation that attracted Crystal Skippers. If the butterflies could succeed there, proper landscaping could promote their persistence over much larger areas.

NEW SCIENCE FOR CONSERVATION

Given the extensive development and presence of invasive species, I at first viewed the conservation problems of the Crystal Skipper as intractable. Allison took a more positive view. She created a plan to improve understanding of the Crystal Skipper's ecology and thus to improve its conservation. She succeeded by drawing on a range of research techniques. She honed her focus on the conservation threat posed by urbanization. If good habitat occurred in people's yards or along beachfronts, could Crystal Skippers move out of the big parks, over houses, and through neighborhoods to reach other conservation areas?

Allison first used traditional approaches to track the butterflies as they moved in and around habitats. She did this by marking the wings of individual butterflies with a unique code using a Sharpie marker. The task may appear straightforward, but it is not. A first challenge is to catch butterflies to mark. I like to find rare things, to engage in a competitive challenge, and to perform repetitive tasks. This places the challenge of butterfly catching and marking among the most enjoyable tasks of my job. The Crystal Skipper presents a formidable challenge in comparison to other rare butterflies that are easy to net, such as the St. Francis' Satyr, the Miami Blue (*Cyclargus thomasi bethunebakeri*), and others. When I saw a Crystal Skipper that was a target to mark, I chased the elusive, small brown butterfly up and down dunes while trying to keep an eye on it. The Crystal Skipper was fast, and it took a steady hand and rapid swing to catch it.

All subsequent stages of the process, from extracting the fragile butterfly from the net to holding it gently to prevent wing damage to marking its wings, were nerve-racking. Even in the net the Crystal Skipper flew vigorously. I wanted to avoid damaging it and at the same time to prevent it from escaping. Once it was caught, I marked each Crystal Skipper with a unique number. Over the following days, our group focused on capturing marked butterflies and recording whether they were on the same or on a different dune than the one on which they had been found when marked. With the resulting data in hand, Allison was able to work backward in her analysis to determine when and where each butterfly had been marked, how long it had lived, and how far it had moved. Most importantly, by noting a butterfly's location when it was marked and when it was recaptured, she could trace a straight-line path between locations. She could then map the parts of each path that traversed dunes, woodlands, and developed areas.

Allison studied habitat areas that were about four hundred yards apart. The number of butterflies that moved between areas separated only by grassy dunes was three times the number that moved between areas separated only by houses. Across the small numbers of butterflies, some flew very short distances and others flew very long distances. Statistically speaking, this produced results that were not significant. Regardless, an important finding for conservation was that the Crystal Skipper was able to move through developed areas. Populations were thus interconnected. If natural or human-caused disturbance extirpates a small population separated by natural or urban areas, other areas can provide a source of colonists. This allows conservation to leverage the power of a metapopulation structure (described in chapter 2). Restoration even in small areas can support new populations of Crystal Skippers.

With data on marked and recaptured butterflies, Allison could accomplish more than just track their dispersal through natural and urban landscapes. She could make use of the data in a more formal analysis to estimate Crystal Skipper population sizes. As Allison often reminds me, this rare butterfly can be locally common. On days in April or June when adults are flying at their highest abundances, one can find a Crystal Skipper readily in the large parks. It helps that few other butterflies live in the harsh environment, so there is little risk of confusing other species for it.

Allison found that the Bear Island population size was approximately six thousand butterflies, and Fort Macon was home to some four thousand. As these two bookends make up the bulk of the range-wide population, it is safe to assume that there are not many more than ten thousand Crystal Skippers. This number serves as a baseline for another goal of conservation and recovery. This involves monitoring whether the population sizes are stable, increasing, or decreasing over time. Protected areas provide a reasonably stable base for the population. We can then monitor whether habitat conservation leads to higher Crystal Skipper numbers.

As her marking studies were limited to assessing dispersal within just a few hundred yards, Allison needed other techniques to understand the connections between populations in parks separated by ten or more miles. She searched for evidence of longer-distance dispersal between populations. This type of dispersal could have happened over many decades, since beachfront development fragmented the Crystal Skipper's range. To implement her search, Allison turned to genetic tools that would allow her to assess population-level changes in DNA over time. If populations were truly isolated from one another, they would show larger differences in gene fragments. Likewise, greater similarity in gene fragments would be indicative of higher rates of dispersal between populations and transfer of genes from one population to the other. Allison took advantage of this science by capturing butterflies, bringing them to the lab, and analyzing similarities in gene fragments across their range.

Allison made a surprising discovery that had new implications for Crystal Skipper conservation. She found no evidence that houses created barriers to Crystal Skipper dispersal over the long term. Instead, natural landscape features such as ocean inlets and long stretches of maritime forest imposed the largest barriers. Taken together with the marking study, Allison's findings pointed in the same direction: Crystal Skipper populations were compatible with at least some beachfront development. This was true as long as good-size areas of butterfly habitat are interspersed throughout the entire extent of the island. Allison and I learned that people could coexist with rare butterflies as long as they do not destroy all available habitat. Allison's scientific advances provided important guidelines for the next steps in Crystal Skipper conservation.

REVERSING PRESENT-DAY THREATS

For a butterfly with small populations that occupy a tiny area of a popular resort destination, the Crystal Skipper has proved remarkably resilient. This is in part thanks to the two bookends of habitat in the state parks. These parks support populations that provide a source of butterflies that can spread to areas that are disturbed. The two bookend populations are not completely safe, however. Even with numbers in the thousands, these populations are still small, and small populations are always at greater risk of extinction. Minor changes in habitat could wipe them out; random changes can reduce reproduction and survival; inbreeding and other genetic changes lower genetic diversity; and catastrophic events such as hurricanes always lurk. The Crystal Skipper has withstood burgeoning beachfront development, stabilized sands, and severe weather. Still, these threats may yet extinguish the Crystal Skipper.

For recovery, the Crystal Skipper's range must increase in area. Through her research, Allison has charted a course of action. On the longest of the two barrier islands that support Crystal Skipper, Bogue Banks, development is extensive and progressing; the only way to increase the butterfly's global population here is to engage people in low-cost conservation and restoration, especially on the dunes in front of their beach houses. This requires planting Seaside Little Bluestem, removing unwanted invasive species, and limiting disturbances people cause by walking on dunes. To accomplish these goals, Allison began to consider the education and restoration interventions required to expand populations between and especially adjacent to where people live.

The largest areas with untapped restoration potential are not between the bookends but beyond them. Crystal Skippers may have once occupied the uninhabited island east of Bogue Banks. Shackleford Banks, part of Cape Lookout National Seashore, is nine miles long and harbors Seaside Little Bluestem. However, Shackleford Banks is also home to about one hundred feral horses. They are descendants of animals introduced by people in the sixteenth century, but how they arrived is not resolved. While not native, these horses receive protection because of their cultural heritage. People love them so dearly that they make great efforts to sustain and expand their populations at the expense of other features of the environment.

Allison conducted experiments by excluding horses from small areas of the island. When she compared those areas to controls, she found that horses selectively grazed Seaside Little Bluestem to nubs, rendering the plants incapable of supporting Crystal Skipper caterpillars.

Horses are likely to remain, yet opportunities for conservation exist. Most immediately, some of the grasslands can be fenced. In lieu of waiting for butterflies to cross the sound from Bogue Banks, Crystal Skippers could be actively introduced to jumpstart their population growth.

To the west of Bear Island, Brown's Island is another target for butterfly range expansion. Brown's Island is on the coastal edge of the Camp Lejeune Marine Corps installation. Marines fire artillery from the ocean over the island to the mainland. Because of live fire overhead, the military prohibits access to Brown's Island, and so there is no record of the Crystal Skipper there. As with the St. Francis' Satyr, the Taylor's Checkerspot (*Euphydryas editha taylori*), and certain other rare butterflies, the Crystal Skipper might find a safe haven on military land.

Based on my observations of other rare butterflies, I find no reason to believe artillery fire harms the Crystal Skipper. It may even help it, as it keeps people out. Yet, a major challenge exists because of limited civilian access: we can't study it firsthand. If Allison or others find the butterflies on Brown's Island, the global number of islands occupied by Crystal Skipper would increase by 50 percent. I have scanned aerial imagery for evidence of high-quality habitat. Shrubs and trees line some stretches of the beach, and these areas are surely unsuitable. Other stretches appear to be dunes covered with grasses and look indistinguishable from the known habitats that the Crystal Skipper occupies. Given the known distribution of Seaside Little Bluestem, it is almost certainly present on Brown's Island.

THE FUTURE IS GRIM

As I look into the future, I struggle to see the Crystal Skipper enduring. This is true even if people adopt the two strategies for maintaining habitat: limiting development on top of dunes and investing in restoration.

People on the barrier island cannot change one threat, sea-level rise. Beachfront property is invariably at low elevation. As a back-of-the envelope calculation, I can project future sea-level rise based on trends over the past century, when the sea level off the North Carolina coast rose by about one foot. This would yield predicted sea levels about a foot higher by the end of the twenty-first century. Recent estimates (2015) of sea-level rise paint a gloomier outlook. The global rate of rise since 1990 has been three times higher than it was in the prior century. These rates are likely to accelerate as temperatures continue to warm, glacial ice melts, and water expands. In likely scenarios of sea level rise in the range of one and a half to three feet, the barrier islands that host the Crystal Skipper will diminish in size. Bear Island is at greater risk than Fort Macon State Park. Even if the world's human population limits increases in atmospheric carbon dioxide and in the rate of sea-level rise, momentum already exists toward loss of these barrier islands and loss of Crystal Skippers.

As sea levels rise, there is one potential course of action to save the Crystal Skipper and other rare butterflies (such as the Miami Blue) living on soon-to-be-submerged barrier islands: move them inland. So-called managed relocation requires that people are willing to take Crystal Skippers from their current range and move them to other places that are ready to receive them. It is not enough to transport a Crystal Skipper to a new place. Even harder will be finding environments, restoring habitats, and establishing Seaside Little Bluestem there. As wave energy is needed to maintain Seaside Little Bluestem's current populations, it is unclear how new environments could support the plant's population. In addition, conservation initiatives must run in parallel with social acceptance of moving rare butterflies. People living in what would become the butterfly's new habitats would have to identify areas for restoration. The process is daunting.

I struggle with proposals to move species, whether the Crystal Skipper or any other species. I am always in search of ways that will work to save rare butterflies in place. The number of species that people could possibly move is trivial. Perhaps people would be willing to move more charismatic mammals, birds, or flowers. They might even be willing to move a few species of butterflies. Efforts for managed relocation, however, will fail to move all the insects, plants, microorganisms, or other species that compose a functioning ecological system. After a decade of ambivalence, I have come to accept that people should move some species threatened with sea-level rise that are trapped on coastal islands. I am skeptical that managed relocation can work for the Crystal Skipper, even though it might work for others (see Miami Blue, chapter 5).

ELEVATING THE BUTTERFLY TO ESCALATE CONSERVATION

Allison completed her dissertation in 2009 and moved on to a position with NASA. In 2015, she received a call from John

Burns. He had finally given the Crystal Skipper a latinized scientific name, *Atrytonopsis quinteri*, its species epithet in honor of Eric Quinter, who first discovered it. Burns had assembled the various strands of information regarding the butterfly's color patterns, anatomy, and habits. In comparison to its closest relatives, the Crystal Skipper differed in its feeding habitats, number of generations per year, and geographic locations. In a more detailed analysis of morphology, Burns found that the Crystal Skipper had more spots, a grayer wing color, and an egg with an orange belt that other species lack. Ultimately, Burns, the world expert on this group of skippers, determined that it is a full species.

Before that decision, the Crystal Skipper's anonymity had provided yet another wake-up call about the unknown branches of the tree of life—unknown branches that can be lost before their discovery. The effort to name the Crystal Skipper took the persistence of the top scientist in this area of taxonomy and of a graduate student who never imagined to be at it for as long as she was. Their efforts yielded a clearer understanding of the diversity of life that is in need of protection.

For now, and before the devastating effects of sea-level rise take hold, Allison found that immediate conservation can occur in front of people's houses. To make that happen will take scientific studies of restoration. It will also require people to embrace restoration where they live. There continue to be halting efforts in this direction, as Allison leads efforts to propagate Seaside Little Bluestem and to empower people to plant it in their yards or other restoration areas. Longer-term conservation and recovery of the Crystal Skipper can occur only after beachside protection and restoration. This process has not taken shape, but I am optimistic that, with surmountable obstacles in the way, it could.

The Crystal Skipper is not the rarest butterfly species. Still, I consider a species that numbers around ten thousand individuals and occupies a small area of narrow islands worthy of consideration. In my ongoing search, I found that stronger contenders have numbers that are lower by several thousands of individuals.

CHAPTER 5

MIAMI BLUE

In 1916, British entomologist George Bethune-Baker took on the task of categorizing the species of one type of butterfly throughout the Caribbean region. The group was a scientific subfamily classified just above a genus and known commonly as the blues. I find species across the entire group to be strikingly similar, and I cringe at the thought of Bethune-Baker's undertaking. As a rule, butterflies in the group have blue upper wings, light gray or white underwings, and gray spots or lines (see, for example, the Fender's Blue, chapter 3, and the British Large Blue, chapter 8). At first glance, a Cassius Blue (*Leptotes cassius*), Ceraunus Blue (*Hemiargus ceraunus*), or Nickerbean Blue (*Cyclargus ammon*) could be mistaken for the Miami Blue (*Cyclargus thomasi bethunebakeri*), with which they co-occur. To distinguish among them when looking at a butterfly through binoculars, I focus intently on spots or lines that differ slightly in their locations on the underwing (the part of the wing that shows when the butterfly's wings are folded behind its back). A key diagnostic that I identify quickly on the Miami Blue is its four black spots (Plate 8). Three are in a line, and only a thin strip of gray separates them from the body; the fourth is offset from the uppermost of these three, toward the outer edge of the wing. Miami Blues also have a distinct white strip near the edge of the hindwing. Differences among the blue butterflies can be subtle, and Bethune-Baker and others of his era did not have genetic tools at their disposal to sort out invisible differences. As he worked to determine how the Caribbean blues related to one another, he found one specimen categorized incorrectly as the Nickerbean Blue, a Cuban species. This was the Miami Blue.

Bethune-Baker did not describe it as a new species. Nearly three decades after his observation, American Museum of Natural History entomologist William Comstock revised the entire group based on wing form and color. He named this new butterfly *Hemiargus ammon bethunebakeri*. Shortly afterward, novelist and lepidopterist Vladimir Nabokov reexamined this and related butterflies by the form of their genitalia (standard practice in butterfly classification). He then gave Miami Blue the unique scientific name that has stuck, *Cyclargus thomasi bethunebakeri*. Entomologist Alexander Klots quipped that "the full scientific name of the Florida race is really something to make even a hardened entomologist wince."

Between the time of its discovery by Bethune-Baker in 1916 and through the 1970s and 1980s, the Miami Blue was common throughout South Florida. It was the butterfly equivalent of a weed, easily spotted against a background of shrubs along roads or around hedgerows. Professor Jaret Daniels of the University of Florida's McGuire Center for Lepidoptera and Biodiversity recounted how in the 1980s he could find Miami Blue caterpillars virtually dripping from the suspected host plant, Nickerbean (*Caesalpinia bonduc*). Butterfly scientists and collectors ignored the Miami Blue, as they were in search of more interesting species.

The Miami Blue received its common name from the city that was the epicenter of its habitat. As recently as fifty years ago, the range of the Miami Blue rimmed the coast of the southern half of Florida, with the top of its U-shaped distribution reaching nearly three hundred miles north of Miami to Daytona Beach on the east coast and to Saint Petersburg on the west coast. Though broad, its distribution centered on Miami and (the very few) points south.

The geography of South Florida is such that species confined to that region are prone to becoming rare. The region is the only part of the contiguous United States that supports tropical moist forest. Located at the very southern tip of a four-hundred-mile-long peninsula, its forests are cut off on three sides by ocean and on the northern side by cooler temperatures. Within this bounded geography, consistently high temperatures throughout the year combine with seasonal rainfall to support a unique set of butterflies. As people encroached from the north, there was nowhere for those butterflies to migrate to. They were stuck in tiny areas, and a small footprint of people created a disproportionately large reduction in the number of butterflies.

When I first traveled by airplane to Miami, as the plane descended from thirty thousand feet, the natural features of the environment drew my eyes. The Atlantic Ocean to the east and the extensive wetlands of the Florida Everglades to the west bound the city. The urban area of Miami and the surrounding sprawl extended more than one hundred miles north to south and covered over one hundred square miles of land. It was not hard to explain the Miami Blue's decline in or near the city. However, its populations extended throughout the entire South Florida peninsula, including to urban and more natural areas. The Miami Blue remained abundant and wide-ranging for much of the twentieth century.

INEXPLICABLE DECLINE

An ongoing mystery in the history of the Miami Blue is why its populations plummeted so rapidly. Beginning in the mid-1970s, its range shrank to marginal areas around South Florida. As recently as 1980, it lived in eight, widely separated sites. These included barrier islands off the eastern and western coasts of Florida, as well as Everglades National Park and the Florida Keys. Even when the Miami Blue appeared safe at those sites, its rapid descent was continuing unabated. The butterfly was last observed on the western side of Sanibel Island in 1990; on the mainland at Matheson Hammock just south of Miami in 1991; on Key Largo in 1991; and on Big Pine Key in 1992. Looking back on this series of coinciding dates, it seems clear that some unknown, rapid change occurred across South Florida in the years prior to 1992. The explanation for this slide toward extinction mystified scientists.

By 1992, only one population of the Miami Blue remained. This last population was on Adams Key off the eastern coast of mainland Florida's southernmost tip. There the Miami Blue faced another threat, hurricanes. On August 24, 1992, one of the most powerful (and to that date the costliest) hurricanes in history, Hurricane Andrew, struck South Florida. The hurricane made landfall at Biscayne National Park, on Elliott Key, a large island just north of Adams Key. Its sustained winds reached as high as 146 miles per hour, and gusts reached 164 miles per hour.

The hurricane's toll was devastating; it destroyed more than 125,000 homes and left about 160,000 people homeless in Miami-Dade County. It caused an estimated $27 billion in damage.

Hurricane Andrew caused the Miami Blue's extinction, or so it was thought at the time. No one knows exactly how the hurricane affected the butterfly. Winds may have swept away adults, knocked caterpillars to the ground, or uprooted the butterfly's host plant. The only other hurricane stronger than Andrew, the Labor Day Hurricane of 1935, caused the apparent extinction of another butterfly, the Schaus' Swallowtail. Given South Florida's extension into such a common hurricane path, it is inevitable that the strongest hurricanes target South Florida's rare butterflies with regularity.

I have often wondered how rare butterflies endure the pounding delivered by such powerful storms. Locally, effects can be devastating. A key to a species' persistence is that other, less affected populations nearby provide an infusion of new butterflies. Butterflies can experience loss in one part of their range while they thrive in others, an emblem of the metapopulation structure I discussed in chapter 2. The Miami Blue's problem on Adams Key was that there were no other populations to repopulate it.

Scientists were still perplexed. They did not know the mechanisms of the species' rapid population decline even before Hurricane Andrew. To explain some of the patterns, Emily Saarinen of New College of Florida tapped into museum records. Specimens have dates and locations pinned to them. Saarinen discovered a striking pattern that helped to explain some aspects of the Miami Blue's distribution and decline. She reconstructed a visual narrative of change to the landscape of South Florida. She divided more than eight hundred museum records into three time segments. The first specimens appeared in the so-called Frontier Era, before 1930. This era encompassed the period of early development of Miami and South Florida as the population grew to about half a million. The majority of Miami Blues captured during this era were in or near Miami.

As the human population of Miami surpassed two million people, it entered its second phase, the Development Era, which lasted until 1970. Records of the butterfly clustered around Miami and extended north and south. As the human population surpassed five million people, Miami entered its final phase, the Globalization Era. Very few specimens occurred in sprawling Miami. Several records were from Florida's southwestern coast, centered on Naples. Most were collected throughout the Keys.

RESURRECTION AND PROPAGATION

On November 29, 1999, seven years following the Hurricane Andrew extinction, the story of the Miami Blue took a surprising turn. Naturalist and butterfly enthusiast Jane Ruffin walked down a path in a small state park, Bahia Honda, located on the key of the same name, thirty-five miles from Key West. She encountered a small blue butterfly that she at first did not recognize. Ruffin knew butterflies well and cycled through possibilities among the many near-indistinguishable blues. No known butterfly fit, causing her to expand her field of view. After studying the butterfly and taking photos of it, Ruffin was convinced that it had to be a Miami Blue. Other experts quickly confirmed her observations. Biologists estimated a population size of about fifty individuals.

One reason for the Miami Blue's persistence on Bahia Honda was the presence of its host plant, Nickerbean, growing readily near roads and trails. Nickerbean is a shrub that can surpass six feet in height. It has many compound leaves, thorns on its stems, and prickly seedpods. On swaths of Bahia Honda, it grows quite well. As Nickerbean occurs elsewhere in South Florida, however, some other features of Bahia Honda's environment essential to the butterfly must have remained intact also. As I will discuss below, scientists are working hard to find these features.

The Miami Blue's small population size and its location within a popular tourist destination exposed it to threats. The small population occupied a small range, making it particularly vulnerable. Jaret Daniels led a group to study the Bahia Honda population. They estimated that Miami Blue habitat extended to about an acre. Bahia Honda is the shape of a long (two-and-a-half-mile) club. The butterfly's habitats were concentrated in two areas on the southern coast. One was on the narrow club handle, where the former path of a century-old railroad cut along a raised berm. The other was on the club head, pinned between a lagoon and the ocean.

Jaret's group quantified Miami Blue population size and trends using standardized surveys. His team walked paths through the butterfly's habitat and tallied the maximum number of flying butterflies. The methods employed by Jaret's group balanced the need for rigor and a sustained effort with the potential harm caused by catching and marking delicate blue butterflies. The minimum possible effort is simply to document the presence of a species at a site. The maximum effort is to conduct a study in which butterflies are captured and marked and then recaptured. The approach can be used to estimate survival and to determine the number of butterflies that are present but not observed. These pieces of information are needed to calculate the most stringent measure of population size. Jaret's group chose a method that produced an index of population size that is well correlated with the more stringent measure.

For nearly a decade after the Miami Blue's 1999 rediscovery at Bahia Honda, Jaret found the population to be stable. Each year, maximum population counts rose to around one hundred butterflies, with seasonal population sizes fluctuating between highs near two hundred and lows near zero. This distinct seasonal pattern through time included months-long periods when the Miami Blue would remain as a caterpillar (Plate 9, top), a time when no adults were present. When conditions were right, Nickerbean growth promoted caterpillar growth, metamorphosis, and emergence as adults. Because some butterflies hid at any life stage (especially as a chrysalis or caterpillar), actual population sizes were in fact higher than two hundred.

Though apparently stable, the small and constricted Miami Blue population was overexposed. A realistic plan for recovery required establishment of new populations on other keys or on the mainland. Obvious starting places for reestablishment were the locations the Miami Blue last occupied, until the early 1990s. As Bahia Honda Key is located three miles from the nearest key and forty miles from the Florida mainland, this could not happen by natural means. Miami Blues fly yards, not miles, and people would have to intervene. Such an approach would mimic efforts underway for the Bay Checkerspot and, as I discuss later, the Large Blue. Unlike those butterflies, the Miami Blue had populations so small no one could sanction removal from Bahia Honda of the number of individuals needed to restore a population elsewhere.

Jaret and his group created an alternative seed stock. They perfected efforts to raise Miami Blues in the lab. They collected a few butterflies at Bahia Honda and brought them into the lab. A female Miami Blue can lay over one hundred eggs. To house the developing butterflies, Jaret's group lined lab tables with dozens of cups. Each contained eggs or caterpillars and a sprig of Nickerbean. There the Miami Blues completed their life cycle until they emerged as adults.

In a boon to the population of Miami Blue, captive adults mated and laid eggs in a seemingly endless cycle. Protected from predators, the Miami Blue could achieve high population growth. By 2009 Jaret's group had raised over thirty thousand butterflies, a number that dwarfed the wild population size. This provided hope that the butterfly could be sustained and given the boost it needed to expand its populations to areas in its former range. The captive population provided insurance against catastrophe at Bahia Honda.

The burgeoning captive colony produced adults to release in the wild. Jaret's team lined up everything needed for reintroduction. They identified suitable sites with butterfly host plants. The sites all occurred within the historic range of Miami Blue, including sites that had supported the butterflies before the wave of extinctions in the early 1990s. Jaret's group narrowed targets to Everglades National Park and Biscayne National Park. In 2004, they released more than twenty-five hundred adult Miami Blues.

If successful, the butterflies would fly, mate, lay eggs, and sustain a population. The reestablished population would then multiply its numbers in its former range. This did not happen. There was no evidence that the butterflies completed a full generation. As the butterfly population did not establish, it was back to the drawing board. Apparently, some essential ingredient was missing. Jaret's group was unable to identify the causes of the failed effort.

I took away two lessons from their experience. Most important, I learned that a successful program requires knowledge of subtle details of the butterfly's natural history. Different life stages may respond in different ways to environmental change, and there are many possible changes. Habitat can be degraded when it is lost, when sprayed with toxic chemicals (such as insecticides), when invasive ants replace ant species that live in partnership with caterpillars (Plate 9), or for other, unknown reasons.

I also learned how these human-caused environmental changes must be interpreted in the context of natural environmental variation. A natural change that I have found to be critical for other rare butterflies is drought. In the case of Jaret's lab-raised population, an intense drought followed the butterfly releases and lasted well into the following summer. Nickerbean experienced severe drought stress, which resulted in compromised food sources for the newly emerged caterpillars. In the late summer and fall, an active season of tropical storms and hurricanes increased moisture and created a water surplus. This prolonged some reproduction and the ongoing presence of the Miami Blue. Without detailed studies, the relative importance of human-caused and natural environmental change in preventing establishment of the Miami Blue is unknown.

RE-REDISCOVERY

As there had been no successful reintroductions by November 2006, Bahia Honda still held the only known population of the Miami Blue. Then, as now, the possibility always existed that other populations of the Miami Blue still lingered at unknown locations.

In the fall of 2006, US Fish and Wildlife Service biologist Tom Wilmers and butterfly enthusiast Paula Cannon embarked on a search for the Miami Blue. They targeted remote islands in the backcountry of the Key West National Wildlife Refuge. On a visit to Boca Grande, a small island located twenty miles west of Key West, they were astonished to discover a Miami Blue population. They then traveled five miles farther west to the Marquesas Keys, where they found six more populations of the Miami Blue in a ring of short, thin islands. During every trip to these islands, spread across the next half year, they found tens to hundreds of butterflies. Populations were larger and occupied a greater area than the one at Bahia Honda. Protected by their remote location, away from most people and the threats they bring, these populations appeared to be thriving.

Over the next five years, however, Miami Blue population sizes at these new locations did not appear as robust. Not a single visit to the islands yielded numbers that reached the high levels Wilmers and Cannon had reported. Most often, numbers were zero, and on rare occasions they rose to ten or twenty. Numbers approaching zero were alarming, especially since no one could explain the population declines in the previous few decades that had befallen the Miami Blue elsewhere. These new populations could easily succumb to the same fate.

Ongoing counts cast doubt on the numbers first reported. Experts were skeptical, because lifelong butterfly scientists had not discovered or measured the populations. Wilmers and Cannon did not have the history of spending hundreds or thousands of days catching and identifying a diversity of butterflies, including the Miami Blue. A plausible hypothesis was that they had confused the Miami Blue with a common butterfly, the Cassius Blue, which looked very similar in appearance and flew on these islands in abundance. Butterfly biologists' inability to see the Miami Blue (at least in quantity) in the new locations gave rise to suspicion and doubt.

RE-EXTINCTION AVERTED

No matter the population size, this discovery came just in time. Progress with the Bahia Honda population had faltered. The population began to decline. Reintroduction had failed. Alarm grew within state and federal agencies, nonprofit organizations, and academic scientists. Then disaster struck—twice.

The first disaster resulted in the end of Jaret's group's captive rearing program. The orderly scientific process that led to successful efforts to breed the Miami Blue was turned upside down, jeopardizing the captive population. Agencies in charge of potential reintroduction sites were nervous about expanding the range of an endangered species to or near private land, where the Endangered Species Act could impose restrictions on landowners. This delayed the permitting process. Without permits, scientists could not release the thousands of captive butterflies into the wild.

At the same time, the biologies of the native and captive populations came into conflict. Apparent declines in the wild population at Bahia Honda caused agencies to impose more restrictions on taking individuals in support of the captive population. As a small number of female butterflies had founded the greenhouse population, the captive population had low genetic diversity and was subject to intense inbreeding. Captive butterflies were less fit and had poorer survival and reproductive capacity, making them less likely to survive both in the greenhouse and when released into natural conditions.

Jaret and other scientists at the McGuire Center began to question the value of their program. With severe restrictions and no clear path forward, they made the controversial decision to let the captive population die out. At the time, this decision made sense, especially in light of the center's impressive ability to raise many more butterflies in the future when circumstances changed and they were able to bring more butterflies in from the wild.

A second disaster ensued when the Bahia Honda population collapsed. Miami Blues were sighted at Bahia Honda in November 2009, in winter 2010, and finally in July 2010. There were no observations afterward, and the Miami Blue was now considered extirpated on Bahia Honda.

Scientists attributed the extirpation to a deadly combination of changes to the physical and natural environment at Bahia Honda. The initial catastrophe was severe, frigid weather that reached unusual lows. Cold air from the jet stream pushed south. On January 11, 2010, Key West experienced its second-lowest temperature, 42 degrees Farenheit, in over 130 years. Cold weather lingered for nearly two weeks. The cold temperatures reduced adult Miami Blue activity and struck at a bad time when numbers were already at a low point. The Miami Blue was not adapted to frigid cold.

Cold weather also harmed the butterflies indirectly by damaging their host plants. The weather killed some Nickerbean leaves and retarded plant growth. Any Miami Blue caterpillars that survived had lower chances of finding food. Yet, this cold snap could not have been the sole cause of the butterfly's decline. Every year there are dry periods when caterpillars wait for rain and for plants to grow. The extreme cold did not kill the Nickerbean plant, which was capable of regrowth. The year prior, drought had knocked back Nickerbean, and it had regrown quickly. After the drought, the plant and butterfly had both shown the capacity to recover.

Other biological changes interacted with physical environmental changes. Green Iguanas (*Iguana iguana*) were now abundant at Bahia Honda. These non-native lizards had arrived in South Florida by stowing away on ships that crossed the Gulf of Mexico and the Caribbean Sea from the West Indies and Central and South America. They exerted their fullest and most consequential effect on the Miami Blue by transforming the food web. These iguanas are mottled green to light brown in color, and they grow up to five feet in length. While exploring Bahia Honda, I found large and small iguanas hanging from trees and shrubs. It was easy for me to flush them from vegetation.

Green Iguanas proved fatal to Miami Blues. As iguanas do not eat caterpillars by choice, they did not target Miami Blues. They are herbivores that eat many different types of plants, including Nickerbean. In a typical year, dense growth of Nickerbean provided plenty of food to sustain both butterflies and iguanas. Under stress, however, leaves died back to the stem and only sprigs of edible new growth remained. In 2009, all the most tender and tasty leaves on many plant species experienced the same fate, either withering or defoliating. The frigid conditions in the Keys caused a food scarcity that limited iguana and butterfly populations. Iguanas consumed the new sprouts of Nickerbean, leaving little for the dwindling populations of caterpillars. Iguanas also ate caterpillars, though only as a side effect of their consumption of leaves. Some considered the iguanas the cause of extirpation. I was skeptical that iguana removal alone would ensure Miami Blue persistence. As with the broader decline of the Miami Blue over the previous decades, no one could be certain about the causes of its disappearance at Bahia Honda. While the loss of the other populations still lacked explanation, it seemed plausible to attribute loss at Bahia Honda to some combination of frigid temperatures, low food availability, and competition with iguanas.

A FRESH PERSPECTIVE FOR THE MIAMI BLUE'S FUTURE

In the aftermath of the loss of the Miami Blue at Bahia Honda, the failed reintroductions, the loss of the captive population, and the discovery in the Marquesas Keys, state and federal agencies reached an impasse about how to proceed with the butterfly's conservation. At their root, disagreements arose because the causes of decline remained uncertain. The US Fish and Wildlife Service then asked me to join the studies of the Miami Blue, with hopes that I might serve as an arbiter. I had research experience with endangered butterflies, though not with the Miami Blue. Perhaps I could bring a fresh scientific perspective. I joined the effort to assess threats to the remaining populations and to propose actions that might improve conservation efforts.

On my first trip to the Florida Keys in September 2010, I traveled with my former graduate student Johnny Wilson. We flew into Miami at night and drove south to Key Largo. As we passed Homestead and began to cross the waters of Long Sound and Manatee Bay, we encountered the leading edge of Tropical Storm Nicole. It had been ill-advised for us to travel to South Florida during the heart of hurricane season. Rain fell all night long. The highest recorded rainfall from that storm was twelve inches on the northern end of Key Largo. We awoke the following day to drizzle and clouds. The weather and roads appeared clear enough to allow us to proceed. We surmised that our original plan to travel farther west into the narrow and exposed arc of the Keys was a bad one. We rerouted and went to view prime habitat for future restoration near the Flamingo Visitors Center in Everglades National Park. The only insects I remember encountering there were dense clouds of mosquitoes obscuring our views of any butterflies. I did my part to feed them.

We found the Nickerbean in the Flamingo area to be tall, lush, and abundant. Fortunately for the Miami Blue, loss of habitat did not mean complete loss of food plants. Nickerbean had been the focal host plant because it proliferated on Bahia Honda. In fact, Miami Blue caterpillars will also feed on other host plants, including Florida Keys Blackbead (*Pithecellobium keyense*) and Balloon Vine (*Cardiospermum corindum*). These different host plants are especially important because Nickerbean is not present on the Marquesas Keys, and Florida Keys Blackbead is. Range-wide, all these plants are found on the Keys and on the mainland, though they have been reduced in abundance by development, particularly on oceanside dunes. They are now interspersed among emerging developments and expanding fields. Given the regularity and ease of finding the host plants, however, I cannot imagine that food loss could be the cause of the Miami Blue's extirpation range-wide.

Once the tropical storm had moved on, Johnny and I worked our way out to the lower Keys. We were too late to see the Miami Blue on Bahia Honda, as it was already extirpated there. We continued west toward locations where the Miami Blue persisted. Arriving late, we spent the first night at a bunkhouse on Big Pine Key. The next morning, Johnny and I drove thirty miles farther southwest to meet US Fish and Wildlife Service biologists Phillip Hughes and Tom Wilmers at a dock in Key West.

We boarded a boat that would take us west to the keys that still harbored the Miami Blue. A flats boat is thin and lightweight, necessary qualities for boating in much of the area surrounding the Keys, where the water is but a few feet deep. Our destination islands were twenty to twenty-five miles away. "Running the flats" represents a balance between speed, to cross long distances of open water, and light weight, to allow the boat to ride high, on top of the water. Wilmers, a seasoned captain, drove us at speeds I had not imagined possible. The half hour trip across open and shallow water took us past a number of small keys. Travel to the islands was a thrill: a fast boat ride and regular encounters with sea turtles and tropical seabirds. The weather was beautiful, with clear blue skies, even if temperatures were roasting hot. Our destination was on the far edge of a flat shelf just before the next ocean channel.

On this first trip, I held both great anticipation and anxiety. For me, the success of our trip hinged on whether we observed the Miami Blue. We arrived at our first stop, Boca Grande, in mid-morning. We hopped from the flats boat into the warm, shallow water. The beach was narrow, and our destination was just beyond where the beach began. After wading just a few feet, we scrambled up a two-foot-high sand dune. Beyond where the dunes degrade into beach, the plant life of the island was at first grassy and herbaceous, interspersed with many flowers. This first part of our walk was easy. Then we hit a near-impenetrable wall of vegetation. Within a few feet, the plants had changed to solid hedges, some draped in thick vines. The shrubs that made up these hedges were important: on these islands, Florida Keys Blackbead served as the primary host plant for the Miami Blue.

Nearly immediately, we found what we had come for. We stood between open fields and dense shrubs, and a small light blue butterfly fluttered along the edge where the habitats met. Compared to other butterflies, it was not a fast or strong flier. Identification nonetheless proved elusive, because of its erratic flight and its similarity to other blue butterflies. Over many trips to these islands, I spent hours observing the colors, patterns, and behaviors of the look-alikes to confirm my ability to identify the Miami Blue. This first one perched on a Florida Keys Blackbead flower. I focused my binoculars directly on it. As far as I could surmise, it was a Miami Blue.

About five minutes later, another blue butterfly flew by. Up for a new challenge and eager to confirm its identity, I ran after it for twenty yards. I failed to keep up. When I observed the next blue butterfly, I deployed my net. With success, I cornered the butterfly in a fold in the netting, lightly pinched its wings, and drew it out for others to see. I transferred it to my other hand, gently wedging its thorax between my fingers, and showed everyone its wings. I'd succeeded in capturing a Miami Blue. To my surprise, Wilmers had never before captured a Miami Blue nor seen one this close. This was a new and memorable experience for all of us.

We did not plan to conduct a formal count of Miami Blues on that first trip. Our objective was to scope out future possibilities for research and conservation. This made our observations of dozens of Miami Blues especially gratifying. I was impressed that we observed them everywhere we searched. Wilmers had not observed numbers this high since he had first discovered the population five years earlier. Our new observations presented a puzzle. Why had numbers dipped to such low levels in the intervening period?

SCIENCE IN THE BACKCOUNTRY

Over the next three years, my lab took on the responsibility of measuring the population numbers of the Miami Blue. The effort to determine the true sizes of Miami Blue populations, as with those of all rare butterflies, can be tedious, time-consuming, and exasperating.

Our greatest initial challenge was to get to the places where the Miami Blue lived on a regular basis. It may appear delightful to cross flat open water in South Florida; however, there are two obstacles. First, traveling to and from the shallow areas where the butterflies live entails crossing deeper channels, where larger waves and winds easily toss a flats boat around. Second, weather in the Keys is erratic and at times dangerous, and afternoon storms can develop rapidly, churning the seas and making for a rough crossing. On not infrequent days of impending bad weather, we canceled our field plans. Even in good weather, we did not cross when the view from the channel's edge was of white-capped waves. This limited the number of days that we could conduct field research.

To estimate the sizes of Miami Blue populations, I brought on a new student, Erica Henry, who had years of experience studying rare butterflies in the Pacific Northwest. She worked between the mid-morning start of predictable butterfly activity and the early afternoon, when winds churned the seas. Add in the tides, which when outgoing pushed back on the return trip, and the resulting window for her work spanned about four hours. She did not have time in a single trip to sample the entire, combined four-mile length of beachside habitat at Boca Grande and the Marquesas. Instead, she developed a regular sampling scheme whereby she identified one point every thirty meters (about one hundred feet). From those points, she counted all the butterflies she saw in one minute and within a circle with an eight-meter (twenty-six-foot) radius. Walking or boating between points, she could sample all areas within a few days. Statistical methods allowed her to scale samples within a small fraction of habitat area to all Miami Blue habitat. Using this approach, she estimated the total Miami Blue population size to be about eight thousand butterflies.

While conducting her studies, Erica encountered the same problem as those before her. At times, she arrived to find the butterflies flying in good numbers. At other times, she saw no butterflies. Active and inactive seasons for adult butterflies were expected. In higher latitudes in North America or Eurasia, butterflies fly all summer and do not fly in winter. In the Keys, Miami Blue numbers fluctuated just as widely, even without a distinct winter season. This left Erica unable to explain why, over three successive years, she observed peak abundances in September, then in March and April, then in the following March, July, and August, and then in July through October. On every trip, Erica faced uncertainty about whether or how many butterflies would be flying. Were they at a low or a high point in their population cycles? When population counts were zero, had she seen the last butterfly?

Erica solved this problem by examining the relationship between Miami Blue numbers and rainfall. Unlike temperature, which tends to be very consistent throughout the year, rainfall varies, with a wet season, beginning as early as June, followed by an extended winter–spring dry season. With little change in temperature, rainfall is the key factor affecting plant growth.

Erica first correlated a given day's rainfall with the number of Miami Blues she saw that day. She found no relationship. One limitation to her methods was that her counts were of adult butterflies. During a butterfly's life cycle, the adult was not the stage that she predicted to be most sensitive to precipitation—it was the caterpillar. Caterpillars, rather than adults, are dependent on lush vegetation to feed and grow, so we would expect a lag in adult abundance caused by the time caterpillars need to eat, grow, and metamorphose.

Erica used what she had learned to predict adult abundances. She searched her data for a delay in adult abundance after rainfall and caterpillar feeding. She considered rainfall in a period beginning two months in the past, then she recorded all the rainfall that accumulated over the next month. Using this amount of rainfall data, Erica discovered that she could accurately predict the number of Miami Blues one month later. She hypothesized that during the dry season, caterpillars suspend their feeding and growth, which is then stimulated by rain.

This science provided crucial information needed to resolve what had been a lingering source of contention in Miami Blue conservation. Wild fluctuations in Miami Blue abundance caused different people to observe very different numbers that ranged from zero to hundreds. That they recorded these numbers at the same locations gave rise to different perspectives. Pessimists saw low numbers as dire warnings. Optimists saw high numbers as signals of resilience. Reconciliation became possible only after the discovery that rainfall patterns explained population fluctuations. The easiest times to navigate to the islands were when there was no rain, during the dry season. Repeated visits during extended dry periods would yield no or few butterflies. An understanding of Miami Blue natural history provided one important guide for conservation and recovery.

THE OTHER CLIMATE CHANGES

Erica's findings refocused the attention of those of us who worry about climate change. Whereas climate models do not predict much change in this region's temperatures, they do predict change in the timing and intensity of rainfall. Hurricanes, fueled by warmer seas, have already become more severe. They bring more rainfall. Of special importance to these small islands, they increase wind and salt spray. More generally, models predict the annual amount of rain to be more variable. Years of sustained drought could devastate butterfly populations. Erica is now using her data to create models to project Miami Blue population sizes under future climate conditions.

Other research has pointed to a catastrophic outcome of climate change for the Miami Blue: sea-level rise. Rising seas eat into the edges of small islands, such as those the Miami Blue inhabits. In her five years of research, Erica observed that stretches of dunes on Boca Grande lost 115 feet of their width. Accelerated sea-level rise will inundate these islands. Given the rise to date and the levels predicted for the future, the Miami Blue is not safe as long as it is limited to these small island environments.

In September 2017, the possibilities of climate change became real. Hurricane Irma, then a category 5 storm, came barreling toward the Florida Keys. As it approached, we glued our eyes to radar images. A direct hit by the eye wall could wipe away Boca Grande and the Marquesas Keys and with them the Miami Blue. Passage of the eye wall just to the west could be just as or more severe. This is the "dirty side" of a hurricane, where its counterclockwise rotation causes winds and water to crash forcefully toward land, causing a powerful storm surge.

In the end, the eye wall struck to the east (the "clean side"; storm surge to the west of the eye wall is less severe). The storm rotated across land, which reduced its wind speeds and potential for destruction. Irma still caused two different types of damage, one degrading and one enhancing Miami Blue habitat (Plate 9, middle and bottom). On Boca Grande, wind speeds were still strong enough that salt spray killed the shrub layer that included large populations of Florida Keys Blackbead. On areas of the

PLATE 1. Top: Nick Haddad in Tikal National Park, Guatemala. Bottom: Mediocre Skipper (*Inglorius mediocris*).

PLATE 2. Bay Checkerspot (*Euphydryas editha bayensis*).

PLATE 3. Top: Bay Checkerspot caterpillar. Bottom: Serpentine grassland at Coyote Ridge, California, home to the Bay Checkerspot's last population.

PLATE 4. Fender's Blue (*Icaricia icarioides fenderi*).

PLATE 5. Top: Fender's Blue caterpillar on Kincaid's Lupine (*Lupinus oreganus*). Bottom: Our research group after one of us found a Fender's Blue caterpillar at Baskett Butte, Oregon; the butterfly's habitat shows in the background.

PLATE 6. Crystal Skippers (*Atrytonopsis quinteri*).

PLATE 7. Top: Crystal Skipper caterpillar on Seaside Little Bluestem (*Schizachyrium littorale*). Bottom: Allison Leidner scans habitat at Fort Macon, North Carolina, site of one of the two largest populations of the Crystal Skipper.

PLATE 8. Miami Blue (*Cyclargus thomasi bethunebakeri*).

PLATE 9. Top: Miami Blue caterpillar (center), tended by ants, on Florida Keys Blackbead (*Pithecellobium keyense*). Middle: Hurricane Irma killed Florida Keys Blackbead, a Miami Blue host plant, on Boca Grande. Bottom: Irma deposited soil that will form the base of new habitat on Main Beach in the Marquesas Keys.

PLATE 10. St. Francis' Satyr (*Neonympha mitchellii francisci*).

PLATE 11. Top: St. Francis' Satyr caterpillar on its host plant, a sedge named *Carex mitchelliana*. Bottom: Artificial dams deployed to restore wetlands.

PLATE 12. Schaus' Swallowtail (*Heraclides aristodemus ponceanus*).

PLATE 13. Top: Schaus' Swallowtail caterpillar. Bottom: Our research group enters hardwood hammock habitat in search of the Schaus' Swallowtail.

PLATE 14. British Large Blue (*Maculinea arion eutyphron*).

PLATE 15. Monarch (*Danaus plexippus*).

PLATE 16. Poweshiek Skipperling (*Oarisma poweshiek*).

Marquesas, storm surge caused sand to wash over the dunes, inundating the grassy habitat. As a result, dune area increased and provided the template on which new plant communities could grow. The Miami Blue escaped catastrophe.

Another rare butterfly, the Bartram's Scrub-Hairstreak, was not so lucky after Irma and provided a harbinger of what is likely to befall the Miami Blue. One of the hairstreak's populations, on Big Pine Key, was in the direct path of Hurricane Irma's eye wall. That key experienced a three-foot-high storm surge. The next spring, dead plants fueled fires, creating another powerful disturbance. The storm provided the most palpable test of the resilience of Bartram's Scrub-Hairstreak (*Strymon acis bartrami*) to disturbance. Although no adult Bartram's Scrub-Hairstreak was observed in the year after Irma, there is rare evidence of caterpillars feeding on Pineland Croton, providing some hope. The hairstreak serves as a model for the Miami Blue when it experiences such a direct hit.

GETTING OFF THE ISLANDS

The Miami Blue's history speaks to its resilience but also to its fragility. It has emerged from apparent extinction. It has clung to small populations in tiny shrub forests on the ocean's edge, exposed to hurricane-force winds and burning sun. It has done this in the face of increasing threats created by people, who have developed its habitats, introduced Green Iguanas, and altered the climate. The seas are on the verge of inundating the entire subspecies.

My most optimistic hope is that Miami Blue populations are lingering elsewhere, in remote places that are difficult to reach. With this in mind, Erica teamed with a University of Florida graduate student, Sarah Steele Cabrera, in 2014, to search for

Miami Blues in unexplored areas. They targeted the backcountry of Great White Heron National Wildlife Refuge, searching to the northeast of Key West on Snipe, Sawyer, and Content Keys. They visited under ideal weather conditions, especially at times when recent precipitation corresponded to levels in Erica's predictive models. They discovered two adult Miami Blues on Snipe Key and possible eggs on the other islands. This added a third set of Miami Blue populations to those at Boca Grande and the Marquesas Keys. It also added to the possibility that more backcountry searching will yield other populations. Observations like these set a hopeful tone to our wishes for the butterfly's resilience.

Yet, stepping back to view its long history of fast decline, we must admit that the Miami Blue is fragile and appears relegated to extinction. This is especially likely if we cannot explain the causes of its rapid decline. Jaret Daniels once told me that in the early 1990s he camped on Adams Key and found every Balloon Vine pod occupied by Miami Blue caterpillars. They were so common he ignored them. Jaret also recounted that at the time, as he worked to recover other rare butterflies, the thought that the Miami Blue would someday need protection was not on his mind. Yet its populations plummeted. Though clearly correlated with the expansion of the human footprint, this decline has been mostly inexplicable.

Any realistic plan to recover the Miami Blue will involve people replacing the natural dispersal process and moving butterflies to restored and protected habitats within its former range on the mainland. Natural disturbances such as hurricanes and shifting sands around the margins of small islands expose the places it lives to degradation. Rising sea levels will wash these islands away within decades. Protection of the existing habitats is a Band-Aid that will offer fleeting protection for the

Miami Blue. Restoration of the butterfly in historic habitats could improve prospects, but there is nearly no chance that the Miami Blue will ever reach those areas, located twenty or more miles across open ocean.

Within the next decade, scientists and agencies must transport the Miami Blue to higher ground in its former range. This is akin to the problem the Miami Blue faced when scientists thought it existed at one isolated and exposed location at Bahia Honda. New efforts are needed to perfect reintroduction techniques. I will note that this effort is not the same as managed relocation, as I discussed for the Crystal Skipper (chapter 4). The Miami Blue once inhabited the mainland and its host plants are already there. In theory, reintroduction should be easier.

New and sustainable populations must be established on the contiguous Keys and then on the mainland. Once they are established, scientists and conservationists can address other threats. The slow march of science has brought this conservation action closer into view. We now know, thanks to Erica's work, how to predict the Miami Blue's abundance in nature. This allows scientists to time the collection of wild individuals for captive rearing when it will least impact natural populations. Individuals raised in captivity can be released to seed new populations.

Early in 2016, I had the opportunity to visit Miami Blue populations on an expedition led by Erica and Sarah. As predicted by recent rainfall (higher because this was an intense El Niño year), many butterflies were flying, and they had laid many eggs. To my eye, finding these eggs was nearly impossible. Food was abundant, but the eggs were minuscule. Erica and Sarah had the uncanny ability to sleuth out many, many eggs. Even when they pointed them out to me, I had trouble seeing them against the background of leaves and stems.

Jaret and Sarah collected eighty-one Miami Blue eggs and caterpillars in November 2016 to use to reinitiate captive rearing. Three months later, they had over one thousand chrysalides and thousands of eggs and caterpillars. This speaks to the inherent ability of butterflies to grow their populations at a rate much higher in a lab setting than in nature. My lab has taken advantage of this same feature of butterflies when we reared the St. Francis' Satyr (chapter 6) in captivity. On a recent trip to South Florida, I traveled from the airport to O'Deeney's Caribbean Restaurant in Florida City to have dinner with Erica and Sarah. Contained within two plastic cups, a thousand Miami Blue caterpillars joined us. These were products of the successful program to propagate them at Jaret's Miami Blue lab. Sarah was en route to release them within apparently suitable habitat at Long Key and Big Pine Key. The goal was to use captive butterflies to establish new wild populations.

Sarah is now working to resolve features of Miami Blue natural history to shine light on its requirements. As I have learned and seen repeatedly, swarms of butterflies can be set free in apparently good habitat, and the introduction effort can fail—and we have no understanding of why. Sarah's efforts will help guide future endeavors to reestablish butterfly populations.

Among other things, Sarah is seeking clues about the contrasting food sources for Miami Blue on Boca Grande and the Marquesas Keys and on the mainland. Many butterflies have two or more host plants. Could the offspring of butterflies raised on Florida Keys Blackbead in the Marquesas survive on Nickerbean at Bahia Honda? On one trip to the Marquesas, Jaret located a lone Nickerbean plant. He found it loaded with Miami Blue eggs and caterpillars. This was the first evidence that the same population of Miami Blue could survive on either plant. As Nickerbean is easier to raise in quantity in a lab setting, Jaret gained new support to use it as the primary food source for captive populations. Sarah tested the performance of caterpillars in adjacent Nickerbean and Florida Keys Blackbead. She placed caterpillars on each plant species and then, to prevent them from leaving, fastened nets around the caterpillars. She reported higher success on Nickerbean. It remains uncertain whether higher survival can vary between environments. For example, perhaps Miami Blue caterpillars perform better on fast-growing Nickerbean when conditions are favorable to the plant, only to do better on Florida Keys Blackbead in harsh conditions that the blackbead can withstand and Nickerbean cannot.

Sarah also asked: Can Miami Blue eggs or caterpillars released onto wild plants survive to become adult butterflies? She introduced caterpillars onto plants that were covered with netting in potential new habitats to monitor all life stages of Miami Blue growth. The next step is to determine whether she finds the same results when she grows caterpillars on unprotected plants or when free-flying adults produce new generations of butterflies.

Is the Miami Blue the rarest butterfly in the world? It would have been if I had created this ranking at a previous time. Coming off historic highs in the 1970s, the butterfly reached catastrophic lows in the 1990s. After resurrection in 1999, it flirted with extinction in 2009. Although it was not known at the time, 2006 was a year Miami Blue achieved a new plateau. In the Marquesas, hundreds of adults were observed on a single day. These observations, when adjusted to account for births and deaths through the generation, translated to an even larger population size. Looming threats to the Miami Blue include the vulnerability imposed by the small area it occupies and the inevitability of the seas rising around it.

The Miami Blue offers a tale of caution. With its once lofty population size, it is an exemplar of the possible rate and depth of population declines. How can a butterfly's fate change so rapidly—with no apparent explanation? It is hard for me to look at any common butterfly now and see it as secure, even the Monarch, as I'll discuss in chapter 9.

There has been real progress in the science and protection of the Miami Blue. Although still low, its numbers—in the thousands—are higher than were expected just a half decade ago. For now, the Miami Blue is twice as abundant as the next rarest species.

CHAPTER 6

ST. FRANCIS' SATYR

Within a year of my arrival as an assistant professor at North Carolina State University, the US Army asked me to advise and apply scientific research to the conservation of the endangered St. Francis' Satyr (*Neonympha mitchellii francisci*). I did not hesitate to become involved, motivated by my interest in conservation and rare things in general. What I did not anticipate were the long hours, days, and years I would sink into the St. Francis' Satyr.

The most surprising aspect of the St. Francis' Satyr's story was the army's central role. The army has been the key institution in this butterfly's discovery, ecology, and recovery. I was at first apprehensive about working with the military on conservation, worried that it would be antagonistic to concerns about an endangered butterfly on its land. In my first year, I learned my worries were misplaced. I met the garrison commander, Colonel Tad Davis. After I delivered a briefing, he shook my hand and said, "We often equate 'serving our country' with combat. I would like to see that expression broadened to include other activities such as promoting environmental stewardship on military training lands." With that statement, I knew I could count on the army's support.

A SOLDIER'S SURPRISING ROLE

The story of the St. Francis' Satyr's discovery requires some context. Viewed from a distance in nature, the St. Francis' Satyr is not large, gaudy, or colorful. It is less than two inches across with its wings spread. It is nondescript, the color of a dead fallen leaf or exposed organic soil. I see subtle beauty in the St. Francis' Satyr that is apparent in a closer view (Plate 10). Its several reddish orange stripes (fading into a brown background) run the length of its wings, which have a series of black dots with yellow outlines and small, embedded silver spots.

Second, people hunting butterflies within its habitat have to distinguish it from several others of the same size and shape. In order from least (but still reasonably close) to most similar, Carolina Satyr (*Hermeuptychia sosybius*), Little Wood Satyr (*Megisto cymela*), Gemmed Satyr (*Cyllopsis gemma*), and Georgia Satyr (*Neonympha areolatus*) live with, or within a stone's throw of, the St. Francis' Satyr. To distinguish these butterflies, one needs to capture them or observe them at close range.

Third, the habitats where the St. Francis' Satyr lives are mucky, grassy wetlands, typically barricaded by shrubs and vines. For this and other reasons, these environments are difficult—at times nearly impossible—to traverse.

Fourth, the butterfly occurs only at the Fort Bragg army installation in southern North Carolina. Within that locale, almost all of its populations are in areas restricted to access, two artillery ranges. This leaves only three groups of people with permission to enter these wetlands—soldiers, army staff including biologists, and members of my lab.

I see soldiers in training at Fort Bragg nearly every day that I am in the field. Their typical days might involve practice missions with a new platoon, hiking through and camping in the woods, firing practice at the artillery ranges, or—central to this story—training in navigation. One of the main navigation courses starts on a hill above a wide creek. There are many possible routes and obstacles. Soldiers navigate to predetermined destinations after passing through several waypoints. Following one of the navigation training paths, soldiers march downhill through open woods and toward a creek. If they navigate correctly, they circumnavigate the wetland and cross the creek on a small gravel roadway. The correct route is the path of least resistance.

These are soldiers in training. Once off course a soldier orients directly through a wetland. Anyone on this route must penetrate a ten-foot-high barrier of shrubs, so tall and dense that it is impossible to see over or through. Although not the intended path, this is the straightest route as the crow flies. This path takes soldiers into a wetland replete with rare plants and animals sensitive to human destruction.

This is an uncommon path that is taken with enough regularity to be noticed. On many days when I walked in the wetlands, I came face-to-face with a soldier soaked in mud to the knees or waist. Even when our trips to these wetlands did not correspond, evidence of the trainees' presence lingered. Habitat damage included lines of parted shrubs, submerged grasses, and boot prints that formed a line through to the other side. Although this type of habitat destruction was limited in its extent, the damage lasted.

On June 2, 1983, Thomas Kral was a soldier on this navigation course who brought with him the skills of a lepidopterist. He is central to three of the most important events in the history of the St. Francis' Satyr: its discovery, its scientific classification, and its listing as an endangered species. Kral slogged through a wetland thicket and followed a small brown butterfly. He must have been skilled indeed to complete a soldier's task while simultaneously recognizing this butterfly as something unique. He had discovered the St. Francis' Satyr.

Shortly after discovering it, Kral recognized that this butterfly was distinct from all others. The next step in the St. Francis' Satyr's story stands in contrast to the plodding process by which the Crystal Skipper received a scientific name; Kral cowrote an article describing the St. Francis' Satyr as a new subspecies in 1989. The process sped along because the butterfly's global distribution was thought to extend to one small population in North Carolina. The other recognized subspecies, Mitchell's Satyr (*Neonympha mitchellii mitchellii*), had been described a century earlier and was found in Michigan, seven hundred miles away.

Differences between the two subspecies were subtle. St. Francis' Satyr females were slightly darker, with a thinner yellow band ringing each of several black spots that line up in a row on the wing and with stripes more rufous than orange in color. Even with photographs of the two butterflies set side by side, I can't detect these differences. Yet, these subtle differences reflect independent histories and geographies that have resulted in distinct evolution.

After the St. Francis' Satyr's discovery, Kral played two roles in its listing as endangered, one positive and one (very) negative. First, he knew from the outset that the subspecies was in need of conservation. He conferred a name "in honor of St.

Francis of Assisi, known for kindness to animals and love of natural beauty," in the hope that it would offer some protection. In his first paper, Kral estimated the number of St. Francis' Satyrs to be less than one hundred individuals. When I read the paper, I found it ironic that in nearly the same breath, Kral and his coauthor reported that they had killed fifty butterflies. These specimens were "needed," so they could measure the butterfly and compare it with other, related species, providing the best available evidence that this was indeed something novel.

Kral's next actions accelerated the St. Francis' Satyr's protection. The butterfly's complicated history took an abrupt turn toward incredulity. Authorities indicted Kral in a landmark case on behalf of the rarest butterflies. Kral and his accomplices had been motivated to collect rare butterflies to stock their personal collections and to reap financial reward. Like stamps, butterflies are more valuable to collectors if they are rarer. What better way to make butterflies rarer than to drive them toward extinction. Before the indictment, I met, by chance, one of Kral's two accomplices, Richard Skalski. I had overlapped with him when I was an undergraduate at Stanford, where he was a pest-control agent. Studies of the Bay Checkerspot and other rare butterflies attracted him.

The case against Kral was damning. The evidence included more than two thousand butterfly specimens, all of high quality, each with exact labeling of the date and location of collection. This hoard included specimens of fourteen of the twenty butterfly species listed as endangered in the United States at the time. Most damning was the extensive written correspondence among the team members, detailing their plots to collect butterflies in restricted areas, to hunt and kill protected species, and to evade law enforcement. One letter was signed "Yours in Mass Murder." Kral wrote: "Because some of the things you sent me are on the Endangered-Species List, I will be careful not to reveal where I got them. . . . It's best to trade 'under-the-table' like this."

In 1993, Kral, Skalski, and a third accomplice, Marc Grinnell, were indicted, and each faced the possibility of prison time plus $250,000 in fines. The outcome was not so severe. The case never made it to trial. In the face of overwhelming evidence against them, all three individuals pled guilty. In comparison to the possible penalty, the ultimate settlement included $3,000 in fines, three years of probation, and hours of community service. Each turned over his butterfly collection to the courts.

EXTINCTION AND RESURRECTION

Another event that elevated the need for protection followed quickly: the St. Francis' Satyr's apparent extinction in 1990. At the time, the subspecies' only known population was in the site of Kral's discovery, an area of about two acres. Because of the butterfly's novelty, collecting pressure was high. Scientists wanted specimens to describe it as a new species, and collectors wanted specimens in their museums and their personal collections. I will describe later how we now know of other factors that were at work, such as the fact that untouched habitat that supports the St. Francis' Satyr can degrade due to natural changes. No one understood the effects of this change on the St. Francis' Satyr in 1990.

In fact, the St. Francis' Satyr was not extinct. Rather, the one population that anyone knew to exist in 1990 was extirpated. But no one had entered the adjacent places that were hardest to search. In 1993, army biologist Erich Hoffman surveyed the entire base, including within artillery ranges where the army typically prohibits entry. On that trip, he discovered nineteen St. Francis' Satyr populations. A more intensive and systematic survey of the installation outside the artillery ranges followed, and he found yet more populations. In total, the St. Francis' Satyr occupied about fifty acres of wetland. The majority of this area was in the artillery ranges.

In 1995, largely because of Kral, extinction, and resurrection, the US Fish and Wildlife Service quickly listed the St. Francis' Satyr as an endangered species. Among other threats, Kral's indictment alerted the US Fish and Wildlife Service to the possible collection of the St. Francis' Satyr nearly to extinction. However, a number of other threats loomed larger.

AN OVER-OPTIMIST

When the army approached me in 2002, they were looking to bring science to the management and protection of the St. Francis' Satyr. They thought my experience in modern field and statistical techniques to estimate population sizes and trends would help achieve recovery goals. I had years of experience studying butterfly populations. What they and I did not consider at the time was that my prior research had been on very common species, such as the Common Buckeye (*Junonia coenia*), Variegated Fritillary (*Euptoieta claudia*), and Spicebush Swallowtail (*Papilio troilus*). Seeking a new challenge, I accepted the opportunity immediately with naive overconfidence.

In my first summer at Fort Bragg, in 2002, my group's efforts were limited to areas outside the artillery ranges (the area inside the ranges remained a black box for some time). My primary goal was to determine whether populations were declining, stable, or growing. To that end, I had three charges. First, my lab was to find areas that harbored additional populations, that could support populations, or that were unsuitable. Second, we were to develop and implement field methods to measure the sizes of the three known populations and any more that we discovered. Third, we were to accumulate knowledge of the butterfly's natural history.

As we began our search, we set our sights on open, grassy wetlands. These wetlands were located along headwater creeks. Most flowed in one channel. Our target was different, broader wetlands where surface water spread over the floodplain and braided streams.

The wetlands we sought were not stable features of the landscape. In an ideal world, each wetland a St. Francis' Satyr population occupied would remain suitable habitat forever. To army biologist Brian Ball's way of thinking, "We need[ed] to figure out a way to freeze St. Francis' Satyr's habitat." This is not possible.

Long before I arrived at Fort Bragg, biologists suspected that some natural disturbance maintained open wetlands. This included some combination of fire—to keep trees and shrubs out—and American Beaver (*Castor canadensis*), which helped spread out water over the floodplain. Both disturbances were important parts of the functioning ecosystem. That first year, however, we did not know which type of disturbance—or with what frequency—benefited the St. Francis' Satyr. Furthermore, butterflies could not survive fire and could not survive flooding. We did not concern ourselves with the habitat dynamics; rather, we searched for the grassy wetlands that existed.

My lab searched every foot of creek or stream on the installation. Going into the search, I was surprised the St. Francis' Satyr had eluded discovery for so long. My view changed completely after just one trip into the riparian forests that bordered possible habitat. These environments were inhospitable.

That summer, we searched a total of forty miles along creeks and streams. Every step was arduous. Before beginning a day's search, I'd strap on a pair of snake boots. These knee-high boots afforded some protection against the one animal that had an uncanny correlation in abundance with the butterfly, the venomous Cottonmouth (*Agkistrodon piscivorus*).

It was easy to arrive in uplands near potential butterfly habitat. Most of Fort Bragg is open Longleaf Pine (*Pinus palustris*) woodland covered sparsely with trees and dominated on the ground by a thick cover of Wiregrass (*Aristida stricta*). The environment was open, with rolling hills. A dense network of unpaved roads crisscrossed the installation. The roads were merely paths of sandy soil made at times more treacherous by large military rigs that left ruts up to two feet deep. The roads brought us within fifty yards of any creek that might harbor the St. Francis' Satyr.

After we stepped out of the car, every hundred yards of the journey seemed like ten times that distance or more. We entered a dense mass of shrubs well over our heads in height. Tag Alder (*Alnus serrulata*), Black Gum (*Nyssa sylvatica*), Redbay (*Persea borbonia*), and other shrubs and young trees grew a fraction of a yard apart from one another, in clusters. Thick layers of vines, especially Roundleaf Greenbrier (*Smilax rotundifolia*) and Muscadine (*Vitis rotundifolia*), connected the shrubs. For the most part, I wove my way through thin gaps in the vegetation. Sharp thorns clawed at me. Where the going was most treacherous I clambered to the tops of bent shrubs, using their force to catapult me forward over other plants. There were times when forest hikes seemed endless.

Although we put in days of fruitless effort, occasionally we reached the ultimate target, a small, grassy wetland no bigger than an acre in size. Abundant grasslike plants called sedges (especially species in the genus *Carex*), a group that includes the St. Francis' Satyr's host plant, covered these wetlands. Interspersed among the sedges were a few trees and clumps of shrubs. Small, braided streams about a yard wide spread through the wetlands.

When I reached a wetland, I focused on two things. The first indication of good butterfly habitat was the presence of the presumed host plants. If I saw them, I turned my attention toward small brown butterflies weaving in and out of grassy cover. Unlike most butterflies, the St. Francis' Satyr does not feed at flowers. Making it even more elusive, its favorite activity was no activity at all. It spent most of its days resting on blades of grass or underneath leaves.

As I trudged across wetlands, I swung a net to brush against plants and scare up butterflies. A day spent carrying a butterfly net can be wonderful, but I found it less so as I slogged through wetlands and their mucky soil. The water rose to just under the soil surface or flowed in sheets just over it. What appeared to be soil could give way as I stepped on it, landing me waist deep in a mud pit. Army biologist Brian Ball once called out to me from behind and pointed down into the footprint I'd just made. I looked to see a Cottonmouth coiled in it. I trudged through wetlands many times on a given stream and then in the dozens of small streams across the army installation.

In the months spent surveying all the base wetlands in that first year, we had discovered three new St. Francis' Satyr populations. These discoveries led me to believe that surely we would find more.

In our second major task, we measured St. Francis' Satyr population sizes. This occupied most of my lab's time. Like many butterflies in warmer areas, the St. Francis' Satyr has more than one generation per year. Adults of the first generation fly for three weeks, disappear for a month as caterpillars mature, and then generation two flies for three more weeks. Conveniently for my schedule and for the schedules of my student technicians, these times spanned the duration of the college summer break. When the butterflies were flying, we visited these wetlands daily.

We surveyed butterflies on transects, dedicated path-like sampling areas through wetlands that standardized our efforts. We quickly discovered a flaw in our methods. If we were to set a course through these wetlands on each visit, we would create a muddy trench, taking down the butterfly's food plants and in effect destroying habitat. A side effect of the trench was the natural reaction to step just outside to the higher ground, effectively widening the trench. We needed an innovative way to cross wetlands without destroying them.

My former graduate student Daniel Kuefler devised a scheme to balance sampling needs against habitat destruction. He created simple boardwalks that traversed each site. They were a series of thin, ten-foot-long planks stretched along a consistent path. The boards raised us (just barely) above the mud and concentrated our foot traffic so we did not stomp on the butterfly's food. The boards were tricky to navigate, and I told my students that by the end of each summer they would be gymnasts. From these boardwalks, we studied the St. Francis' Satyr. Although the boardwalks were part of a makeshift plan in 2002, they have remained ever since.

We initially tracked St. Francis' Satyr population sizes with the most rigorous technique, one that involved marking butterflies. After netting a butterfly, we used a Sharpie pen to write an alphanumeric code unique to each butterfly. We then recorded the code of any butterfly that had been marked. We repeated this task daily, recording the marked butterflies we recaptured, and marking all others. By tracking the loss of marked butterflies, we measured one of the two most important aspects of demography, the daily rate of survival (the other is fecundity). We found that adult life spans were unimaginably short, on average just three or four days. If marked butterflies reappeared after disappearing for a day or more, we could determine how often we missed seeing butterflies even if they were there. We used these two pieces of information in combination to calculate butterfly population size.

In 2002, we estimated the size of St. Francis' Satyr populations outside artillery ranges to be about five hundred butterflies in total. The next year we recorded eight hundred butterflies, and the following year eighteen hundred, the highest number on record. After 2004, the populations steadily declined.

In 2005, I became concerned that we were causing declines in butterfly population sizes through our marking efforts. In the best case and most usual scenario, a marked butterfly will lose a few wing scales but will retain full wing function. The St. Francis' Satyr is a more delicate butterfly than most. The marking could possibly change a butterfly's behavior or make it more visible to predators or more unattractive to potential mates. I could never be sure about the ultimate effect of marking. I suspended our efforts to mark butterflies, even though I learned later that our concerns were misplaced.

In lieu of marking, I turned to a second method. We counted all butterflies on a dedicated path. We walked these paths every day and counted every individual butterfly within sixteen feet of the path. With these data in hand, we used a standard formula that integrated numbers in each generation. By this method, we assumed that we never counted the same butterfly twice within a day and that we observed all butterflies within a thirty-foot-wide transect. I knew that this method would produce less precise population size estimates. To allay my concerns, I conducted an analysis and found that the estimates produced by marking and by counting were highly correlated.

Our third objective was to accumulate knowledge of the St. Francis' Satyr's natural history. The unknown aspects of basic butterfly biology initially surprised me. For example, a crucial missing detail was St. Francis' Satyr's host plant. Without that knowledge, habitat management could not begin. As a starting point, we invested much effort into searching for caterpillars. Despite hundreds of hours of searching in 2002 and for years afterward, we found none.

It took us fifteen years to find just two caterpillars in the wild. Both times, we found them by dumb luck. An undergraduate student, Ross Pilotte, found the first in 2013. One day when he was walking through the St. Francis' Satyr's habitat, he dropped his sunglasses into a large sedge. When he reached to pick them up, his hand extended next to a sedge blade. There on the leaf beside it was a St. Francis' Satyr caterpillar. In 2015, a team of students walked along our boardwalks, brushing against overgrowth while keeping their feet out of the mud. Another undergraduate, Ben Pluer, looked down to keep his feet on the narrow boards. He saw a caterpillar fall onto the board. It was the same color and size as a St. Francis' Satyr caterpillar. It lucked out: it had a one in one hundred chance of falling onto the board rather than into the water. We raised it to determine that it was a larval St. Francis' Satyr.

The St. Francis' Satyr is unlike other butterflies. If adult female St. Francis' Satyrs behaved like other butterflies, they would lay eggs in obvious places where our group could follow caterpillars over time. They would exhibit characteristic fluttering behavior as they lay eggs. In fifteen years, we have observed only a few females lay eggs. St. Francis' Satyrs are also difficult to survey because they sometimes lay eggs on plants that neighbor suspected host plants rather than on the host plants directly.

PRIDE GOETH BEFORE THE FALL

I began my research confident that my efforts to bring science to the conservation of the St. Francis' Satyr would stabilize or even grow the subspecies' numbers. While quantifying butterfly numbers, I assumed that keeping people out of wetlands and letting nature take its course would restore conditions and benefit butterfly populations. My immediate action toward recovery was to protect the butterfly's habitat, which had degraded over centuries. My group kept people and activities out of the St. Francis' Satyr's wetland habitats. As the military was trying to maintain natural environments for training, wetlands were more secure on the installation than in most places.

I then observed instances when the very disturbances that I felt were important to maintain populations actually caused their extirpation. In 2005, a beaver's dam flooded out one population. In 2006, a large wildfire jumped from upland forests into a wetland and incinerated another population. In other cases, the lack of disturbance was responsible for population loss. In 2008, 2009, and 2010, wetlands dried out, and trees and shrubs grew in, replacing sedges and extirpating the St. Francis' Satyr.

It slowly sunk in that both disturbance and lack of disturbance hemmed in the St. Francis' Satyr. Both extremes caused extirpation. I realized the importance of striking the right balance between disturbed and undisturbed sites.

In 2011, there was just one population remaining outside the artillery impact areas. It supported fewer than one hundred butterflies. I had signed on to this research project to put the St. Francis' Satyr on a path to recovery. My optimism had been misguided. We found no new populations after the first year, and fifteen more years had elapsed. The last population was sliding toward extirpation. I had failed. I wondered when I would see the last butterfly outside the artillery ranges.

A BIOLOGIST'S DEATH MARCH

Even as the butterfly declined, the army gave me access to the artillery impact areas. Because of daily bombardment with heavy artillery fired from around the periphery, the army usually strictly prohibited access to these areas. I was fortunate that the army allowed me in.

The artillery ranges were refuges for the St. Francis' Satyr. What was it about the ranges that made them ideal habitat? Two hypotheses came immediately to mind. First, the ranges kept people out. Artillery does not approach the negative effects people cause with intense land uses such as agriculture and urbanization. Second, guns, flares, and other artillery set fires each year, simulating natural disturbance. The only way to test these and other hypotheses was to visit the ranges.

I was now able to enter the heart of the worldwide population of the St. Francis' Satyr. Daily, soldiers lined the dozens of firing ranges that ringed the impact area and fired toward the center. Howitzers lobbed artillery from longer distances. All the firepower was directed into the areas where I would walk.

On my first visit, army biologist Brian Ball and I waited for live fire to cease. After we received the green light, we drove past the firing ranges until we arrived close to wetlands. From there we walked. Few roads penetrated the core of these areas. We were now in remote areas where there were no people. Our destinations were largely predetermined. Biologists had identified some populations in prior surveys. Others had used data collected remotely, for example from satellite or aerial imagery, to pick out likely habitat based on known features of the environment such as geography, hydrology, and vegetation.

A third person accompanied us. Our visit was dependent on an expert in explosive ordnance disposal (EOD). While we searched for butterflies, our EOD expert searched for artillery shells, looking especially for signs that they lay undetonated. Our EOD expert that first day, Tracy Dice Johnson, was all business and surely annoyed that our reason for navigating artillery ranges was a butterfly. Tracy and I seemed to get excited at exactly the opposite times—she when we passed a "forty-mike-mike" (a 40-millimeter grenade, ideally the safe color, blue, and not a dangerous "golden Easter egg") or a more impressive 155-millimeter diameter shell; I when we saw a rare butterfly, plant, or wetland. Especially when shells were large, I was simultaneously awed and frightened. Tracy directed us to areas where artillery shells were infrequent; I do not once remember being told that a piece was live.

Over time, I watched Tracy transform. The first sign was when she, leading the way, asked, "Is that one of your butterflies?" The answer was yes! At first indifferent, Tracy became among the most eager supporters of the butterfly, enabling science and conservation that would be impossible without her.

There is a great paradox about artillery ranges. When I first walked in, I expected to see a moonscape. The ranges were just the opposite. The integrity of the forests and wetlands, including the rare plants and animals, was the highest I had seen in North

Carolina. Within the ranges, disturbance caused by fire maintained open forest. Projectiles, flares, and machine-gun bullets ignited fire and simulated natural disturbance that people exclude where they live. Because of frequent fire, the longest lengths of our path cut through grassy savanna shaded by just a few trees.

We walked at least ten miles in search of butterflies and their habitats. We followed close to streams in search of wetlands. As the floodplains along streams were cumbersome to navigate, we followed the path of least resistance along uphill slopes. Our journey was long, but it never failed to include hidden treasures. We saw endangered Red-cockaded Woodpeckers (*Leuconotopicus borealis*), whose populations were recovering throughout the army installation. As we descended toward streams or wetlands, we'd walk down wet hillsides known as seeps. Water that falls onto higher ground passes out of the sandy soil on these hillsides. Because seeps are constantly wet, they support plant life that I rarely saw elsewhere. At times, we stood in or even on populations of endangered plants, including Rough-leaved Loosestrife (*Lysimachia asperulifolia*) and American Chaffseed (*Schwalbea americana*). We passed large stands of tall Yellow Pitcher Plants (*Sarracenia flava*), more occasional smaller Purple and Sweet Pitcher Plants (*Sarracenia purpurea, Sarracenia rubra*), and even Venus Flytraps (*Dionaea muscipula*), which impressed on me that North Carolina is a center of the world's diversity of carnivorous plants.

When I first walked into a wetland that supported the St. Francis' Satyr, my jaw dropped. The wetland was recognizably different from any outside the ranges. It was large and open. The St. Francis' Satyr populations were dense. Our short visit constrained my ability to estimate population size, but I could count the butterflies I saw. I observed about thirty butterflies in a short, twenty-minute visit to that first wetland, which was no more than an acre in size. I observed similar numbers in five wetlands we visited that day. In contrast to what was happening outside the ranges, these populations were thriving.

Since that first visit, we arranged visits on one weekend during each of the two adult generations per summer. In midwinter, we worked with the army to set dates to access the ranges. We did our best to predict when butterflies would be most active. We had experience to draw upon. Using past data, my research assistant Heather Cayton could compute precisely the day we would see the first butterfly, which set the population in motion. This day was predictable with knowledge of the total heat input at the surface, calculated roughly as the difference between the warmest and coolest temperatures each day, added up over all days in the year to date. For the St. Francis' Satyr, Heather could get us to within one day of its first appearance in nature. This prediction is accurate a month or two in advance, yet we were asked to set dates with the army a half year in advance.

In a good year, we visited eight wetlands. In the decades before I first arrived, others observed five additional populations in the dangerous interior realm of the artillery range, where the army would not grant me access. Taken together, thirteen or so small wetlands located inside the artillery range made up nearly all of the butterfly's global range.

KILLING BUTTERFLIES TO SAVE BUTTERFLIES

After repeated visits to the artillery ranges, my concept of restoration for the St. Francis' Satyr turned upside down. Intense and frequent disturbances caused by artillery fire were not, after all, harmful to butterflies. Large St. Francis' Satyr populations in the ranges pointed to exactly the opposite conclusion. In addition to fire, ranges supported healthy populations of beavers, which flooded large stretches of streams. There was a giant difference between what I observed outside the artillery ranges and what I observed inside. The extensive metapopulations within formed an interconnected network of wetlands, such that individual butterflies from undisturbed wetlands could colonize disturbed areas where the vegetation was recovering. I could see now that disturbance was not harmful to the butterfly populations.

This experience forced me to admit that nearly every measure I had taken to conserve the St. Francis' Satyr outside the artillery ranges had been wrong. My lab's conservation actions had protected the butterflies from direct harm, including via natural disturbances. My rationale had been that the St. Francis' Satyr had existed in these areas for this long, and so, I assumed, it would continue to remain safe.

One of these wrong decisions occurred in my first year of research. A beaver dam grew downstream of a newly discovered St. Francis' Satyr population. Biologists were worried that this dam would grow so big that it would flood out the population. The next year, the beavers were gone. The road engineers had used our concerns and the dam's flooding of a road as justification to remove the beavers. In the ensuing years, the butterfly population boomed. It expanded to colonize the area that had been underwater. A decade later, however, with no further disturbance, the population collapsed.

By doing nothing, I was, in fact, killing the butterflies. If protected from disturbance but otherwise let be, their open habitats continued through a natural course of succession, giving way to shrubs and then trees. These woodlands excluded the St. Francis' Satyr's host plant, and thus the St. Francis' Satyr.

Trees then introduced environmental feedback, acting like straws that sucked water from the soil. In doing so, they caused wetlands to dry. This reduced the habitat's potential for the St. Francis' Satyr even further. The irony was that by preventing fire or floods from killing the butterflies, we were killing them anyway. Herein lies a great paradox exposed vividly in the artillery ranges: we need to kill some butterflies to save butterflies—indeed, to save a species.

Unlike my restoration targets outside the artillery ranges, the environments inside the ranges were turbulent. Fires and floods introduced constant change to the St. Francis' Satyr's habitats. The processes that seemingly destroyed habitat at the same time initiated a process that regenerated host plants and habitat the butterfly needed to thrive. What I observed firsthand was how artillery helped to recreate natural disturbances and habitat conditions that once defined this landscape before the army—before people—ever walked on it. It did so in at least two ways.

First, artillery-generated fires replaced natural fires. Before people occupied the region, wildfires defined the landscape and its forests. Fire scars on tree rings and other evidence indicate that fires burned across the area every one to three years. Historically, lightning ignited fires when striking the senescent tufts of grasses that covered the forest understory; the fire then spread from one tuft to the next as it moved across the landscape.

In this context, the artillery ranges, in which people cause fires that replace natural fires, are among the rare places that have the appearance of natural habitat. Seen from above, the ranges formed a doughnut shape. From the outside edges, the artillery is pointed inward, toward the center of the range. Artillery shells landing in the doughnut's hole, flares shooting overhead, and—most commonly—machine-gun fire aimed within the doughnut ignite fires. The artillery causes annual fires to burn through the ranges. It is on these fires that the St. Francis' Satyr depends.

In some years, especially when conditions were dry, these fires have burned through the wetlands. The St. Francis' Satyr cannot survive fire. Yet, those fires retarded growth of trees and shrubs and maintained open, grassy environments. This situation differed from that outside the artillery ranges. There, controlled fires were set in upland forest each year. For historical reasons, however, these fires did not burn through wetland vegetation. Dense forests along creeks and streams served as barriers for fire, containing it. The legacy of this management technique was that fire could not penetrate the wetland vegetation. A battle I am currently waging is to convince foresters to abandon traditionally held views and to restore natural conditions by opening up dense forests within wetlands to fire.

A second way that artillery ranges conserve the St. Francis' Satyr is that they support healthy populations of beavers. There, the animals have refuge from people. Outside the ranges, people remove beavers after they become pests when they flood roads or fields. Inside the ranges, there are no people, and beavers are able to behave freely.

The St. Francis' Satyr is unique among butterflies in its dependence on beavers. Catastrophic habitat loss for the St. Francis' Satyr culminated a century ago, once people had exterminated all beavers in North Carolina and in much of the eastern seaboard of the United States. It is no small coincidence for the St. Francis' Satyr that the army established its base in eastern North Carolina simultaneously with the extirpation of beavers.

Beavers have rebounded and are now abundant in North Carolina, and they do not appear to limit St. Francis' Satyr populations. Compared to other beaver ponds, however, those I saw within the ranges were different. The number and extent of wetlands affected by beavers impressed me.

Ecologists describe beavers as ecosystem engineers. The great changes beavers and their dams cause to physical environments create conditions required by a diversity of plants and animals. Eventually, beavers change the ecosystem in ways that favor the St. Francis' Satyr. Flooding at first favors aquatic animals and plants. Simultaneously, it diminishes or eliminates species intolerant of flooding, including many shrubs, vines, and trees. Sedges thrive in environments that are wet, with standing water or saturated soils, and where there is abundant light. They move in shortly after beaver ponds recede.

Outside the artillery ranges, I am involved in ongoing efforts to strike a balance between the needs of butterflies, the needs of beavers, and the needs of people. Over his career as a biologist at Fort Bragg, Brian Ball has seen how "people are slowly starting to figure out the genius of beavers." St. Francis' Satyrs do not live in flooded habitats, and they do not live in mature forest. Rather, they live in places beavers have recently been active but have since abandoned. As flooding subsides, it leaves behind the sunny, wet, and mucky conditions that are ideal for sedges. Abandoned wetlands look like grassy fields with shrubs and trees scattered sparsely throughout. Over decades, repeated flooding and abandonment by beavers maintains conditions appropriate for the butterfly.

Butterflies cannot survive flooding. Butterflies cannot survive fire. Or so I thought. Inside the artillery ranges, there is evidence of extensive, persistent fire and flood. These disturbances are in or around all the places the St. Francis' Satyr lives.

Butterfly habitat there is constantly changing, created and then destroyed. The St. Francis' Satyr thrives. My visits to the impact areas transformed my approach to St. Francis' Satyr conservation and to conservation in general.

A CAVEAT

If beavers flooded or fires burned *every* wetland inhabited by the St. Francis' Satyr, the butterfly would become extinct. In the proper balance, some sites must be disturbed while others are not. After disturbance, individuals from undisturbed populations can send immigrants to new ones. This is a classic example of the dynamics of a metapopulation. Disturbance forces extirpation. At the same time, dispersal reseeds populations that had been disturbed but can grow again. My group has been honing our methods to strike the right balance in restoration. In the end, we must kill some butterflies to save all butterflies.

It would be nice to have a rule of thumb for disturbance of St. Francis' Satyr populations. I imagine something like what Cheryl Schultz developed for the Fender's Blue when she found that one-third of the grasslands should burn every year to conserve the butterfly's habitat. The problem is more challenging for the St. Francis' Satyr because there are two types of disturbance. Settling on the relative frequency of the two is a new and pressing area of research for the St. Francis' Satyr.

TAKING OFF THE KID GLOVES

For more than a decade, my restoration efforts had been too passive. I had been assuming nature would take its own course to provide the habitat the St. Francis' Satyr needed. A hands-off strategy had caused the habitat to degrade rapidly. We needed to reverse course, actively and decisively creating disturbances to sustain the butterfly's habitat.

We could have restored the needed disturbances in any one of three ways. The first, and the one I do not like to suggest too loudly, is perhaps the most obvious: we could have asked the army to turn the artillery outward from the ranges into other areas at Fort Bragg. This would never happen, as it would disrupt other training and other environmental management.

Second, natural resource managers could have quickly restored forested wetlands to grassy wetlands by burning them. To do this, they would have worked against the historical treatment of wetlands as barriers to impede the spread of fire. Fires have always been contained to prevent them from escaping the army installation and burning fields, farms, and houses. I foresaw a time in the distant future when wetlands would be burned as regularly as the upland forest, currently every two to three years to mimic historical rates. The time was too distant to promote St. Francis' Satyr conservation now.

Having eliminated two options, I proceeded with a third: I had to train my students to be beavers. To do this, my lab conducted landscape experiments to test how we could best mimic beaver effects. Beavers create wetlands in two ways; they create temporary dams that flood wetlands, and they clear trees and thereby reduce water loss. To test the effects of these activities on St. Francis' Satyr populations, we created an experiment with four treatments: one with a dam, one with tree and shrub removal, one with both dam and tree/shrub removal, and a control left untouched.

For the dam treatment, we discovered temporary damming structures that allowed us to create wetlands for the St. Francis' Satyr that, at least in the short term, were better than beaver dams. We obtained dams from Aquadam, a company that constructs them from plastic tube structures (Plate 11, bottom). These dams can be massive, ten and even twenty feet high, to create reservoirs or levees. This gives some context as to why, when I called to place an order, the salesperson referred to my purchases as "baby dams." Our dams were 150 feet long. For each, we pumped water from a stream into the plastic tube and inflated it to three feet wide and two feet high. The water provided volume as well as the weight needed to keep the dam in place and to resist the weight of water that built up from the stream behind it. Our goal was to elevate the water next to the dam to the level that created standing water. The water level would diminish away from the dam, reaching soil level within about thirty feet. At distances greater than this, soil remained saturated. Simultaneously, the dams spread water out to create wide and braided streams. Without completely inundating large areas, the water killed plants that did not tolerate flooding and provided the damp conditions needed for the butterfly's host plant to grow.

Also like beavers, we removed trees. Beavers open the forest canopy, letting light through to the wetland surface. Incidentally, they remove a source of rapid water loss. The net effect is to increase wetland habitat for the St. Francis' Satyr. Without the aid of beavers, and without the aid of heavy machinery that the government prohibits in wetlands, we resorted to other means to create our experimental treatment.

My position as a professor conferred a great advantage, as an endless reservoir of student energy surrounded me at the university. Those students rose nearly to the skill and energy levels of beavers. In the winters, when leaves were off the trees, professionals felled trees with chainsaws and chopped them into three-foot-long logs. In the spring, teams of students hauled log after log by hand from the wetland into the nearby forest.

Although the work was muddy and tiring, we were eventually able to clear twenty wetland areas, each a quarter acre in area.

Our experimental treatments resulted in wet, muddy pits. Many types of disturbance-adapted plants grew quickly. Before long, dark mud gave way to lush, grassy wetlands. The areas soon harbored the communities of plants required by St. Francis' Satyrs.

Less than a year after we began the experiment, the St. Francis' Satyr colonized our treatment areas, particularly those in which we had removed trees. We were surprised by how quickly the butterflies arrived. At one experimental area, we had stacked the decks in favor of St. Francis' Satyr colonization. We located this treatment area just two hundred yards from an existing population. Although this was not a long distance, we were unsure whether it was in the range of St. Francis' Satyr dispersal within just one year. We have tracked the butterfly population here every day through the summer, each year since the first. Although butterfly numbers were never high at this site, the population has persisted.

A different experimental site was far from any existing St. Francis' Satyr population. This site had special meaning, as it was the site of the St. Francis' Satyr's discovery. Soon after discovery, some combination of people and habitat degradation extirpated the population. To test the effects of our experimental treatment, and with no source of butterflies nearby, we had to add butterflies. We replaced the natural process of dispersal by adding St. Francis' Satyrs raised in a greenhouse.

In parallel to our field experiments, we developed an operation to raise captive butterflies from egg to adult. Raising butterflies entails a number of well-defined steps; if any one of them failed, the captive population collapsed. Roughly, the steps are no different from those followed by children who bring caterpillars in from the wild, feed them regularly, watch them turn into a chrysalis, wait for them to emerge as an adult, and then release them back into the wild.

I never thought this process would limit our experiment. It seemed simple enough. But the St. Francis' Satyr introduced several complications. First, as we almost never saw caterpillars in the wild, we turned to their mothers. After mating, female butterflies have one goal in their short, three-day adult life: to lay their eggs. We surmised that if we could capture female St. Francis' Satyrs and place them in netted cages, they would lay some or all of their fifty-plus eggs. Females were not difficult to catch. We were limited by the number we felt comfortable removing from the small populations. The one place where we felt that populations were large enough to withstand removal of any females was in the artillery ranges.

We did our best to arrange visits to artillery ranges to correspond with the time adult St. Francis' Satyrs were flying. Even then, we often ran into problems. Brian Ball once said, simply, "Just like at a bar, the males show up early and go home late." As a rule, butterfly males emerge first and anticipate the emergence of females. On our trips into the artillery ranges, there were times when all the St. Francis' Satyrs we observed were male. On a good trip, I returned to the greenhouse with three females.

To produce the numbers of butterflies needed for the experiment, my lab coaxed captive females to lay eggs. The setup was simple: we placed each butterfly in a medium-size planting pot. To coax the butterflies to lay eggs, we pulled a few long, skinny leaves from the host plant, wrapped the cut end in wet gauze, and placed them at the bottom of the pot. We covered the pot with window screening. We waited a day or two. Most of the time a butterfly did one thing: sit. Even in nature, the St. Francis'

Satyr is a sedentary butterfly. The butterfly must move to lay eggs. We introduced incentives to move, including fans that mimicked wind and mist sprayers that mimicked rain, and we gently shook the pots to agitate the butterflies. On average, captive females laid about twenty eggs in the two days we held them.

Once we had eggs in hand, the process ran more smoothly. The eggs hatched in about a week. We carefully placed newly emerged caterpillars on host plants. This was a delicate task with a caterpillar that is less than a quarter inch long and a strand of hair in width. As long as the caterpillars had food, they ate and grew (Plate 11, top). We constructed elaborate facilities to ensure sufficient food: we had plants growing in pools of water, mimicking their natural wetlands, with a covering to prevent caterpillars from falling into the water and nets over the plants to keep the caterpillars in and predators, primarily small spiders, out. After about a month, the caterpillars became chrysalides. A week after that, adult butterflies emerged.

We now had the butterflies we needed for our experiment. There was one last obstacle that I wished we could have overcome to increase our capacity for experiments and for restoration. We were unable to induce captive butterflies to mate. If we could just get them to do that, our greenhouse population would have produced butterflies for generation after generation. These butterflies would have been capable of producing many more butterflies than can possibly survive in the wild. Even after we adopted some of the same techniques used in successful rearing of other rare butterflies such as the Miami Blue and the Schaus' Swallowtail, my lab has struggled to solve this puzzle. So far, we have observed captive mating only four times.

With captive-raised butterflies in hand, my lab began to release butterflies in the experimental sites in 2012. Then we waited. If these releases worked, new butterflies would appear months later. We distinguished the butterflies we released from those we did not by marking them. Any new individuals that we observed would be present because of our captive releases and our experimental treatments. When we found unmarked butterflies, we confirmed that the butterfly completed its life cycle in our experimental sites. Each year afterward, we released twenty to thirty butterflies. A few captive-released individuals might generate ten times that many butterflies in the next generation.

In successive years following our first release, we observed 11 wild butterflies in 2012, 106 in 2013, 175 in 2014, 617 in 2015, and 751 in 2016. At this point, nearly all the St. Francis' Satyrs we observed outside the artillery ranges were in the restored area.

I learned a number of lessons from our restoration efforts. First, the majority of new individuals occurred in sites where we had removed hardwoods, regardless of whether we had placed a dam there. Second, for St. Francis' Satyr restoration to succeed, we had to release captive-raised individuals in successive years. Done once, the process would have failed. Third, habitat restoration via disturbance had a limited time window of effect. In just a few years, plant succession to shrubs and trees reduced habitat quality. My lab is currently working to determine the optimal intensity and frequency of efforts. Fourth, I have been surprised that I have not observed negative effects of inbreeding. Just three to five females produced the butterflies we released in the first year and in each successive year. The consequence was that the 751 butterflies in our restoration area in 2016 were descendants of at most thirty females. Each spring I held my breath, dreading the possibility of population decline. Likely because we have pushed population levels so high, I have not seen inbreeding depression so far, and my concerns have diminished over time. Finally, growth cannot continue forever. With any natural population, there are natural limits to population size. One factor that could limit population size is the availability of resources. This does not appear likely for the St. Francis' Satyr in high-quality habitat where the host plant grows densely. Another factor is the presence of predators or disease. We have observed selected cases of predation by dragonflies (especially females of the Eastern Pondhawk, *Erythemis simplicicollis*) and the Yellow Garden, or Zipper, Spider (*Argiope aurantia*). When its populations are high, the St. Francis' Satyr can become a target for predators. In the best-case scenario, the population will remain large and stable. As the population size we saw in recent years was high by historic standards, it is possible that the size overshot its target and must come down to sustainable levels.

True recovery of the St. Francis' Satyr will require new populations and new metapopulations to establish off the army base. There are great opportunities in conservation lands nearby. This will require working with land trusts or other conservation organizations to identify and restore lands. We would then create wetlands with the restoration techniques we have already developed. If the host plant was not already present, we would start with restoration of the plant community. When the conditions are right, we will be in a position to release the St. Francis' Satyr. Finally, after fifteen years, I may understand enough about the natural history of the St. Francis' Satyr and the dynamics of its environment to see this type of expansion work.

Is the St. Francis' Satyr the rarest butterfly in the world? It certainly is competitive for that title. St. Francis' Satyr's global range is one army installation. Within that area, it is restricted to a couple of small artillery ranges. Within those ranges, it occurs in just a few small wetlands. Its close affinity to disturbed wetlands has caused its global range to be restricted to a maximum possible extent, with restoration effort, of about one hundred acres on one army installation. By any standard, this butterfly is among the rarest.

My best back-of-the-envelope calculation of population size started with the seven hundred or so individuals that we saw outside the artillery ranges. Inside the ranges, I derived the numbers from counts of individual populations that I made during about four visits to the ranges each year. Of the eight sites we visited, three or four had substantial numbers, on par with the number in our restoration area. There may have been a small number of populations within the most highly restricted areas of the ranges. Taken together, a reasonable estimate of St. Francis' Satyrs population size is just a few thousand butterflies. By a different measure, the butterfly is rarer still. Even in the artillery ranges, I saw it within twenty acres of wetlands. These areas are for the most part off limits to people. A butterfly with a small population occupying a small and inaccessible range must rank among the world's rarest butterflies.

CHAPTER 7

SCHAUS' SWALLOWTAIL

Miami was not always the sprawling coastal metropolis it is today. Until 1900, the area harbored a few hundred to a few thousand people. This was less than one one-thousandth of the number found in the metropolitan area today. Increased settlement and the onset of development coincided with the discovery of one of the area's native butterflies, the Schaus' Swallowtail (*Heraclides aristodemus ponceanus*). Based on an individual butterfly collected in 1898, entomologist William Schaus published an article in 1911 to announce the subspecies' discovery. The Schaus' Swallowtail is large, similar in size to a Monarch. Yellow wing bands stand out against a black background (Plate 12). The final line of the short article described its habitat. It accomplished this in just one word: "Miami." As the butterfly's range hewed to Miami more closely than the Miami Blue's, its population declined much more rapidly as the metropolitan area grew.

The Schaus' Swallowtail occupies an endangered ecosystem called hardwood hammock (Plate 13, bottom). Hammocks grow on limestone layered with humid soils needed to support tree growth. They lie downhill from more exposed and well-drained limestone soils that form another habitat called pine rockland. The rocklands provide the foundation for much of Miami. Hammocks lie uphill from swamps and inland of the mangroves found on seaward edges. They cannot tolerate inundation by brackish waters. Hammocks support a great diversity of trees, including Gumbo Limbo (*Bursera simaruba*), Paradise Tree (*Simarouba glauca*), and Florida Strangler Fig (*Ficus aurea*), which form a dense canopy overhead. The behaviors of the Schaus' Swallowtail are ideally suited to forested hammocks. It flies fast through openings and weaves through dense forest.

Schaus' Swallowtails typically live one generation per year. Adults fly in late spring and early summer. Their host plants are Sea Torchwood (*Amyris elemifera*) and Wild Lime (*Zanthoxylum fagara*). The caterpillars (Plate 13, top) perform especially well on adolescent plants. Plants that are too young and small will not sustain ravenous, later-stage caterpillars. Those that are too old and mature carry leaves that are tough and of poor quality. Ideal host plants grow just at the edge of hammocks. They burst with fresh, growing, high-quality leaves. Adult butterflies prefer to fly in a mix of sun and shade in places where they can place eggs on plants with new and vigorous growth.

The Schaus' Swallowtail's ideal environment occurs in the aftermath of disturbance caused by hurricanes. Disturbance creates the mix of environments to which the butterfly is best suited. Trees and shrubs that have been broken, toppled, or removed provide conditions for host plant regrowth and flower production by Guava (*Psidium guajava*), Cheese Shrub (*Morinda royoc*), and other plants. Of course, hurricane-caused disturbances can be too severe, as we saw with the Miami Blue. When the eye wall of a hurricane strikes land by or near hammocks, its winds can denude the trees and a severe storm surge can inundate them. Some hammocks occur at elevations just a few yards higher than the lower dune habitat occupied by the Miami Blue, conferring some protection against inundation during storm surges. Hurricanes that make landfall at some distance from hammocks can provide the Schaus' Swallowtail more benefit than harm. As with other butterflies, disturbance can have positive effects on Schaus' Swallowtail populations.

In addition to their effects on habitat quality, hurricanes and heavy rainfall can change the Schaus' Swallowtail's life cycle. The Schaus' Swallowtail is adapted to wait to emerge from its chrysalis until higher rainfall increases the production of its caterpillar's host plant. When conditions are too dry, it can delay by a year or more its emergence as an adult. This takes on particular relevance in the seasonal rainfall of South Florida. Each year a wet season of about five months follows a dry season. Extended wet weather in spring causes a boom in butterfly numbers. Rains brought by hurricanes in fall can increase the likelihood that Schaus' Swallowtails will complete their generation within one year. Extended drought harms the subspecies' host plant, diminishing its availability when butterflies emerge. The ability to delay emergence in the driest years is a tremendous aid to the butterfly's survival in this variable environment.

One side effect of the irregular emergence of the Schaus' Swallowtail is of particular importance to this book. It introduces a complication in estimating butterfly numbers by counting adults alone. Observed butterfly abundances will be high in wet years and low in dry years. These differences are unpredictable and reduce the accuracy of annual counts. Low numbers could be a result of true population decline or because of poor weather conditions. For example, University of Louisville biologist Charles Covell observed low numbers of Schaus' Swallowtails in the mid-1970s that he attributed to drought.

A PLACE OF SORROW

Discovery of the Schaus' Swallowtail coincided with the onset of explosive development of the city of Miami. In the first half of the twentieth century, the metropolitan area grew rapidly to accommodate people who moved from colder and more northerly climates to oceanside properties in warm, subtropical Miami. Development expanded to many hardwood hammock habitats. Shortly after its discovery, the Schaus' Swallowtail was nearly lost. In the early 1920s, development drove the Schaus' Swallowtail out of Miami. The last observation on mainland Florida was in 1924 in Coconut Grove, a small town and now a neighborhood annexed by the growing city of Miami. At that time, there were no records of the Schaus' Swallowtail outside of Miami. Therefore, loss on the mainland equated with extinction.

The Schaus' Swallowtail then resurrected for the first time. Soon after its disappearance from the mainland, it resurfaced in the Florida Keys. In the decade that followed its discovery in the Keys, Schaus' Swallowtails appeared among specimens collected from the eastern third of the hundred-mile-long island chain. Two important locations were Key Largo, just south of Miami, and Lower Matecumbe Key, thirty miles to the west.

Following its rediscovery, disaster struck again. In 1935, a natural disturbance, rather than a human-caused one, pounded the Schaus' Swallowtail's habitat. The 1935 Labor Day Hurricane, still the most intense Atlantic hurricane to make landfall in recorded history, struck at the western edge of the butterfly's range, devastating small Lower Matecumbe Key. Photos of the island post-hurricane show it flattened, including an iconic photo of the Overseas Railroad train blown and toppled off its tracks (Figure 7.1).

In 1940, Florence Grimshawe rang the Schaus' Swallowtail's death knell. She and her partner actively caught and raised the butterfly. Her article "Place of Sorrow: The World's Rarest Butterfly" foretold my inclusion of the Schaus' Swallowtail in this book. She wrote, "Matecumbe, in the language of the long-vanished Caloosa Indians, means 'the place of sorrow.' To the lepidopterist, this name of this Key, which is as rich in legend and mystery as the great fastness of the Everglades, may well have a special meaning—the sorrow of the extinction of a species." After describing her encounters with the butterfly, Grimshawe concluded: "[The Schaus' Swallowtail] made its last stand, when the sand dune and beautiful hammock melted away under the fury of hurricane and tidal wave."

In 1938, butterfly enthusiast William Henderson investigated why the Schaus' Swallowtail had disappeared. To find out, he sought to assemble a complete historical list of recorded dates and locations of Schaus' Swallowtails. He began his quest while working under the assumption that there had been no records since 1935. His approach was straightforward. He tallied all the Schaus' Swallowtails housed in museums and private collections.

Butterflies in collections are particularly convenient for this purpose, as a pin holds the butterfly and a tag that records the time and place of collection. This is just the information Henderson needed. He published his first list, tallying twenty-four

FIGURE 7.1. Train on Lower Matecumbe Key, Florida, blown off its tracks by the Labor Day Hurricane of 1935.

butterflies collected since the butterfly's namesake, William Schaus, had captured the first one. The list served as an advertisement to other lepidopterists who held additional records. Readers responded to the article by sending information contained within their private collections. Over the next seven years, Henderson generated three lists that taken together reached seventy-eight butterfly specimens. Of those, five had been collected in Miami before 1925. In the aftermath of purported extinction, the list contained three findings that were surprising: first, five individuals were from Key Largo and had been collected between 1940 and 1943; second, most were from Lower Matecumbe Key and collected between 1935 and 1945; and third, Florence Grimshawe was responsible for sixty-six of the individuals in collections. Apparently, Grimshawe's eulogy was premature.

Although not extinct, the Schaus' Swallowtail was never common. Writing in a newsletter to butterfly collectors and scientists in 1955, zoologist Frank Young of Indiana University directed readers to butterfly collecting hot spots. When his Florida roundup reached Lower Matecumbe Key, Young advised: "You will probably want to stop here to look for [the Schaus' Swallowtail]. Don't worry about . . . conservation too much. You will probably be happy if you SEE even one specimen." Nearly simultaneously, Alexander Klots of the American Museum of Natural History warned: "Overcollecting [of the Schaus' Swallowtail] by 'game hog' collectors has reduced its numbers seriously. . . . I believe that most people have enough sportsmanship to help protect the species and to refuse to buy specimens at any price." These descriptions spoke to an era when butterfly collecting remained a significant pastime and conservation was still in its incipient stages.

NEGLECTED TO PROTECTED

The Schaus' Swallowtail saga receded for a few decades. Then science and recovery took center stage.

In the 1970s, science and conservation of the Schaus' Swallowtail accelerated. This period marked the first attempts to enumerate the size of the Schaus' Swallowtail population in standardized ways. Before this, records in collections had determined the size and range of its population. Now butterfly biologists would drive or boat to the keys in the butterfly's range and spend days counting the number of individuals they observed. In 1970, Frank Rutkowski observed thirty-five Schaus' Swallowtails on Key Largo. In 1972, University of South Florida biologist Larry Brown counted one hundred Schaus' Swallowtails per day on islands to the northeast of Key Largo in Biscayne

National Park, and University of Louisville biologist Charles Covell counted fifteen Schaus' Swallowtails in Biscayne National Park and several more on Key Largo. These reports confirmed that the Schaus' Swallowtail was not extinct. Yet, its numbers in the wild were dwindling on these small, remote islands.

Two camps began to take shape regarding prospects for Schaus' Swallowtail populations. To my surprise, one camp remained optimistic about the Schaus' Swallowtail's fate. In 1973, Covell and George Rawson summarized a widely held view that "Schaus' Swallowtail seems safe from real or imagined threats of extinction via development, pesticides and overcollection. On Key Largo, developers do pose some threat, but probably not for some years to come." At that time, Schaus' Swallowtails occurred on protected lands, apparently at a safe distance from threats that would reduce butterfly populations further still. Given this, did they need stricter protection? It must have been a surprise when "some years to come" turned out to be a very, very small number. My impression is that the unexpected ingredient was the furious development of the Keys beginning in the mid-1970s.

As efforts to understand the Schaus' Swallowtail's biology grew, another camp began to take shape. Conservationists registered mounting concern about declining numbers and viability. They sought protection for the butterfly akin to that received by mammals, birds, and fish. At that time, however, strict protection for the Schaus' Swallowtail or any other invertebrate animal seemed unimaginable. Yet, the environmental movement was growing, emboldened by the US Endangered Species Act of 1973.

Butterfly biologists vocalized some ambivalence or opposition to government protection. They worried about restrictions on their studies and collections. Frank Rutkowski feared that the Schaus' Swallowtail was being threatened and depleted by developers and butterfly collectors. He also voiced the concern that populations could be reduced inadvertently by overprotective conservationists. This second point may seem like a non sequitur, but it reflected concern that overprotective conservation of habitat would actually harm the butterfly's habitat. This prescient observation followed from the habitat's dependence on natural disturbances for its renewal. Similar concerns apply to other rare butterflies that require natural disturbance, including the St. Francis' Satyr, the Fender's Blue, and the British Large Blue.

Stricter conservation prevailed. After low population counts in the early 1970s, the US government listed the Schaus' Swallowtail as a threatened species under the US Endangered Species Act on April 28, 1976. This was the first butterfly and the very first insect placed on the US list of threatened and endangered species. Recognition of the Schaus' Swallowtail in this way was a watershed moment and ushered in a sea change for butterfly conservation. Within a month of this historic event, the list expanded to include six more butterflies, all from California. In the United States and elsewhere, butterfly protection expanded. This stimulated scientists and agencies to invest more time and money to document population sizes and trends of the rarest butterflies. Quantitative data became a yardstick with which to judge the rarest species and to assess growing threats.

I asked Jaret Daniels, director of the McGuire Center for Lepidoptera and Biodiversity, why the Schaus' Swallowtail was protected first. He noted its large and showy appearance, its brush with extinction in the 1930s, and other threats in addition to accelerating development of the Florida Keys. With these features, it could stand up to scrutiny from those who opposed expanding the list of regulated rare species, especially insects.

Conservation action for the Schaus' Swallowtail cemented an ascendancy of policy created for butterflies. Butterfly biologist Robert Pyle noted: "In the past half century, the preservation of Lepidoptera and their habitats has risen from relative obscurity to become one of the most active subdisciplines in modern conservation biology." This popular attention and legal recognition brought new energy to the science of Schaus' Swallowtail populations.

Sadly, Schaus' Swallowtail populations sustained their ongoing decline even after they received more protection. In 1984, William Loftus and James Kushlan, biologists for the National Park Service, reported very low numbers of twelve or fewer adult butterflies per year in 1979–81. Numbers were slightly higher in 1982, when the biologists observed more than thirty butterflies. They attributed low numbers in 1981 to a dry winter that had retarded plant growth and possibly harmed caterpillars, and higher numbers observed in 1982 to a wet spring and faster growth of the butterfly's host plant. Schaus' Swallowtail numbers were still so low in 1982 that new questions arose about whether the Schaus' Swallowtail's status as a threatened species was sufficient. Maybe its risk of extinction was so high that it should be an endangered species.

The Schaus' Swallowtail tenaciously held its ground in some parts of the Florida Keys just offshore of Miami. Until recently, it appeared that the rate of human development there remained too slow to threaten the Schaus' Swallowtail. The Keys were long enough and the hardwood hammocks covered just enough territory to offer refuge for the butterfly. As late as 1977, Charles Covell believed that, "Biscayne National Monument is large

FIGURE 7.2. Cudjoe Key, Florida, cleared for development, ca. 1975.

enough and has the necessary environmental conditions to maintain [the Schaus' Swallowtail] indefinitely."

The fortunes of the Schaus' Swallowtail quickly turned. The Schaus' Swallowtail's habitat occurs on higher ground, in the same flat, coastal areas where people develop homes to stay above the surrounding seas. The growing human population marched southward from Miami and expanded quickly and forcefully toward development of the Keys. Today, along the hundred-mile stretch from Key Largo to Key West, the development is near complete. Beachfront and inland vacation homes, boating infrastructure, lodgings, restaurants, and other developments extend the entire length. Flip Schulke, who took aerial photos in the early 1970s, recorded the change in dramatic fashion. The tiny islands were mostly white (Figure 7.2). Developers had denuded them of plants to make way for home construction. The ground, a thin limestone layer, was interdigitated with carved out canals that formed a transportation network for boats. Virtually every property connected to the ocean. The same pattern of development was evident in photo after photo, key after key. The loss of habitat was frightening, and it set the stage for the challenge and limits of future conservation and recovery.

CORRELATION IS NOT CAUSATION

It is worth stepping back to consider what caused the Schaus' Swallowtail to decline to such low levels in the century since its discovery. The loss of hardwood hammock was extensive. Its rapid loss followed the growth of Miami starting in the early 1900s. Habitat loss was—and still remains—the primary threat to the Schaus' Swallowtail and the most formidable obstacle to its recovery.

Even so, I find it a good process to reevaluate possible causes. It would be easy to pin all the blame on habitat loss. That would be a mistake. Now that most habitat has been lost, a singular focus might divert conservation attention away from other important threats. Many environmental changes besides habitat loss co-occurred with increasing human population size. Besides habitat loss, the most frequently invoked threats to the Schaus' Swallowtail include habitat fragmentation, climate change, pesticide use, and overcollecting. Some threats have been around for most of the past century, whereas others have emerged as human populations have expanded throughout the butterfly's range.

Did habitat loss, increased use of pesticides, invasive species, climate change, and/or other threats cause population declines? Because all of these threats have grown in concert, it is hard to isolate only one. How does each change affect Schaus' Swallowtail's numbers? Do females lay fewer eggs? Are caterpillars or adults less likely to survive? We cannot be certain of the cause until we have identified the mechanisms that underpin change. This understanding typically emerges from experiments. Because of the Schaus' Swallowtail's small range size, small population size, and federal protection, experiments have been very difficult to initiate.

Habitat fragmentation—changes to the configuration of habitats that remain—happened in conjunction with habitat loss. Hardwood hammocks were sliced and diced into tiny parks or other protected areas. The Schaus' Swallowtails living in these areas were isolated from one another. The areas between fragments, unsuitable to the Schaus' Swallowtail, created barriers to interchange between populations. This fragmentation is in sharp contrast to the interconnected populations that once extended from Miami to and throughout the Keys.

Once the habitat was fragmented, problems arose when a small butterfly population in one isolated habitat performed poorly. Such small populations that lived in an environment with erratic conditions occasionally died out. These areas were now at such great distances from one another that butterflies could not navigate from one population or habitat to another. No one area could provide new colonists to repopulate another. Because nearby populations no longer existed, the Schaus' Swallowtail lived in a disconnected metapopulation. Loss of one population was no longer balanced by gain of another. The endgame of this process is extinction.

Hurricanes can reduce Schaus' Swallowtail populations by their force when they make landfall. The butterfly's range is in a common path of the eyes of gathering storms. Heavy rain, forceful winds, and flooding, including storm surge, can inundate the butterfly's host plants with damaging salt water. Scientists have known that hurricanes can be a catastrophic force for the Schaus' Swallowtail since 1940, when Florence Grimshawe reported its apparent demise due to the strongest Atlantic hurricane to make landfall in Florida. Hurricanes could reduce populations directly through wind and rain and indirectly through disturbance of their food plants. The small population size of the Schaus' Swallowtail reduces its potential to recover from catastrophic hurricanes. This is especially so because of habitat fragmentation, as there is no capacity of populations outside the hurricane's range of damage to provide a source of individuals to repopulate areas that have become extirpated.

Catastrophic hurricanes have increased in frequency and intensity. This change results from air masses that travel over warm summer oceans and generate storms. Climate change warms the air and the oceans in ways that facilitate hurricane development. By increasing the energy available to storm formation, warming temperatures have increased the strength of hurricanes as measured by their wind speed. Stronger hurricanes strike land with more force and have the potential to cause significantly more damage.

Hurricanes have both positive and negative effects on the Schaus' Swallowtail. The US Fish and Wildlife Service linked Hurricane Andrew, which struck in 1992—a direct hit to the butterfly's habitats—to the defoliation of trees in hardwood hammocks and reduction in Schaus' Swallowtail population sizes the following year. Hurricanes that are more distant can have just the opposite effect. Landfall of Hurricane Irma in 2017 occurred on a key one hundred miles west of the Schaus' Swallowtail's current range on Elliott Key and North Key Largo. At the eye wall, there was storm surge of greater than three feet. On Elliott Key, winds were still strong enough to damage vegetation, but this had the effect of improving habitat quality for the Schaus' Swallowtail as regrowth provided new sources of high-quality host plant. In 2018, after Irma, the butterfly population was higher than it had been in 2017, before the hurricane.

KNOWN THREATS WITH UNKNOWN CONSEQUENCES

In addition to known threats such as climate change and habitat fragmentation, invisible threats reduced Schaus' Swallowtail numbers. Even habitats that appeared to have the appropriate host plants, temperature ranges, precipitation levels, and other environmental features did not support the Schaus' Swallowtail.

A likely threat that fell into this category was insecticides used to exterminate mosquitoes. Their use became a serious concern in the 1970s. Schaus' Swallowtail populations continued to decline even after protection and restoration of some hardwood hammocks. There was bound to be collateral damage from the tens of thousands of gallons of general insecticide sprayed annually across South Florida. In most places in South Florida, humans have done an impressive job of knocking back mosquitoes and the diseases they carry. Insecticides target insects generally, and those spread over wetlands could easily drift short distances to hardwood hammocks. The same toxins that harmed mosquitoes undoubtedly harmed other insects, including butterflies.

To what extent did insecticides harm the Schaus' Swallowtail? The best evidence is indirect, gleaned from studies of other swallowtail butterflies. In experimental settings, researchers applied pesticides at levels considered "extremely toxic" or "highly toxic" to caterpillars or their plants to test caterpillar fates. These tests showed consistent negative effects of insecticides on butterfly behavior and survival. In addition to considering the financial cost and the human health value placed on extermination of mosquitoes, we must ask: What is the acceptable cost of insecticide use for the rest of nature?

The potential harm caused by chemicals targeted at mosquitoes led to actions that eventually limited spraying in the Schaus' Swallowtail's habitats. Regulations and federal protection for the Schaus' Swallowtail increased, prohibiting insecticide application directly into hardwood hammock habitat. This did not eliminate the problem. Insecticides sprayed in residential areas, near Key Largo for example, could drift into protected areas. Mosquito control accelerated before modern techniques such as ultra-low volume and GPS-targeted spraying became possible. Population counts provided some support for regulations when Schaus' Swallowtail numbers showed limited recovery in the late 1980s and early 1990s.

Putting insecticide application in the context of other environmental changes that were degrading Schaus' Swallowtail habitat, Jaret Daniels lamented that it was unfortunate that increased mosquito spraying corresponded so closely with butterfly population decline. It generated animosity between those interested in reducing mosquito numbers and those concerned with increasing populations of the Schaus' Swallowtail. This conflict spilled over into other areas of environmental protection beyond the Schaus' Swallowtail. Jaret observed that it took years to move to a new understanding, in which the two sides could work together to benefit the environment and people.

Non-insecticidal methods of mosquito control are possible. Genetic modification of mosquitoes might reduce or even eliminate the need for insecticides to control them. One approach to modification is to insert genes that reduce the abundance and/or health risks of mosquitoes. When successful, a gene passes preferentially (in sexually reproducing species, a greater than 50 percent chance) from parents to offspring—so-called gene drive. Technologies exploiting gene drive are still in development; one prominent application increases infertility in mosquitoes, and another increases resistance to agents of disease. I participated in a program at North Carolina State University that evaluated the costs and benefits of this approach. The topic generated unusually high debate in the Keys, as the first proposed US release of engineered mosquitoes was in Key West to control dengue. A positive (in my view) justification for genetic modification of mosquitoes is the potential environmental benefit of reducing or eliminating chemical control while protecting human health.

Another persistent threat to the Schaus' Swallowtail since its discovery has been overcollecting in the wild. As far as scientists know, this butterfly has always occurred at low abundance and in a limited range. For a butterfly collector, it represents a cherished specimen. In the first half century after its discovery, knowledge of the Schaus' Swallowtail could come only from specimens in collections. The information gleaned has been invaluable in piecing together the butterfly's early distribution and abundance. Yet, just a few decades after the Schaus' Swallowtail's discovery, overcollecting became a threat. There is an insidious corollary to Florence Grimshawe's reports of extinction following the Labor Day Hurricane of 1935. The Grimshawes raised Schaus' Swallowtails for sale to collectors. One suspicion was that their declaration of extinction served to boost interest, sales, and revenue from their operation.

Collecting raised red flags. When, in 1976, the Schaus' Swallowtail became the first insect listed as threatened, collecting it became prohibited. The US Fish and Wildlife Service expressed deep concern about the threat posed by collectors, including reports that specimens were selling for $150 each. The agency concluded that collecting was the most serious threat to the Schaus' Swallowtail. I don't subscribe to this view. For scientists and for the public, collecting has its place. Yet, indiscriminate collecting or discriminate collecting targeted specifically at rare species for personal and commercial gain does not. Despite collecting's central place in the Schaus' Swallowtail's history, the relative impact on its population decline remains unknown. Over the past century, other global changes have been the greatest threats.

NOWHERE TO GO

Government agencies and conservation organizations devoted efforts to curtail the most obvious threats to the Schaus' Swallowtail. They protected the small areas of hardwood hammock that remained. They eliminated insecticide application and reduced it in areas nearby. Yet, these efforts appeared at best to stop the decline and at worst to be inadequate. The Schaus' Swallowtail slipped further away.

To stave off extinctions, scientists turned to the tool of last resort, captive rearing. Scientists at the McGuire Center expended significant thought and energy to raising Schaus' Swallowtails from egg to adult in the lab. They did so with some success. The process of rearing, restoration, and recovery exposed the fragility and the resilience of this swallowtail.

As efforts to rear the Schaus' Swallowtail in captivity advanced, some scientists expressed pessimism. Biologist Bob Pyle warned in 1976 that when considering any butterfly, not just the Schaus' Swallowtail, "The introduction of live individuals as a conservation measure is extreme, potentially dangerous and not to be undertaken without careful consideration of possible effects. . . . The appealing idea of breeding insects for release in the wild is likely to fail . . . [and] introductions are likely to be futile unless conditions are just right." For the next two decades, these predictions proved accurate.

Nonetheless, captive breeding and release of the Schaus' Swallowtail became a central focus of the butterfly's restoration and recovery. Reflecting on his group's decades-long efforts to raise and reintroduce Schaus' Swallowtail, Jaret Daniels told me that "intervention doesn't guarantee survival, but it does offer hope and puts us into a better position to save this species."

As the urgency of the butterfly's need for conservation increased in the mid-1990s, Jaret's group perfected techniques to raise the Schaus' Swallowtail. Scientists at the McGuire Center raised sufficient numbers of Schaus' Swallowtails to proceed with their recovery plan. In 1995–97, they released butterflies at eight locations in the subspecies' present and historic range. Following these releases, scientists identified signs of initial success. Introduction of both chrysalides and adults resulted in observations of adults and mating activity. Apparently, Schaus' Swallowtail numbers had stabilized in the wild, and release of captive-reared butterflies was no longer needed.

These short-term successes did not translate into sustained populations of Schaus' Swallowtails. Newly seeded populations lasted for just a few years before they failed. Why this occurred remains unclear. Initial successes diverted attention away from efforts to study and monitor introduced populations. As a result, no one understands the fate of individuals and the cause of the population loss. Scientists speculated that they had released the butterflies into habitats located too close to residential communities sprayed for mosquitoes. Introduced Schaus' Swallowtails may have been collateral damage.

One lesson I drew from this sequence was that immediate attention must be devoted to follow-up field studies of the butterflies at all life stages. It would have been helpful to study adult butterflies' initial behaviors after release. Did they remain on or near their habitat? As for caterpillars, where did they forage? What were the prospects for survival of adults or caterpillars? Release experiments can answer these questions. Successes and failures can guide future efforts. Ultimately, a successful program would yield observations of sustained populations that no longer require the release of new butterflies. Such efforts may seem costly. In reality, they cost a pittance compared to the costs of captive rearing and on-the-ground conservation. Scientific studies after release of captive-raised butterflies provide a good opportunity to solidify recovery efforts.

Since the early 1990s, biologists have tracked the size of Schaus' Swallowtail populations in more standardized ways. From year to year, population sizes have fluctuated widely. In 1993, populations were low. They rose in the mid-1990s to more than five hundred butterflies, before dropping back in the late 1990s and early 2000s to numbers between one hundred and two hundred.

After 2010, numbers became dangerously low. From 2011 to 2013, the numbers of Schaus' Swallowtails observed in nature were forty-one, four, and thirty-two. These numbers were so close to zero that the slightest disturbance or environmental change could cause extinction. I can't think of another plant or animal that has bounced back from a number so low. Wildlife managers and scientists were now in crisis mode. They established emergency procedures in 2012 to collect females observed in the wild, coax them to lay eggs in captivity, and then raise caterpillars to adults for release into the wild. The fateful search for Schaus' Swallowtail females was now on.

In 2013, Jaret Daniels' group collected two females and seven caterpillars from the wild. This was a game changer. The females laid one hundred eggs, and seventy of those individuals became caterpillars, chrysalides, and then adults. These could then breed in the greenhouse, generating yet more eggs and increasing the captive population for release and recovery further still. I first saw the Schaus' Swallowtails in Jaret's lab in 2015. I expected to see individual caterpillars in individual, secured chambers that regulated their environment. I was surprised to see the caterpillars out on a table, like any other common caterpillar one might rear. My visit left me encouraged that, despite the subspecies' low numbers in the wild, this group of caterpillars could seed suitable habitats and grow natural populations.

Recent releases have proved more successful than efforts from two decades ago. In spring 2014, Jaret and his group traveled to Elliott Key with a number of caterpillars and chrysalides. They planned a significant release to boost the size of a tiny population. Before releasing these new butterflies, they surveyed the area once more. They wanted to know the year's population size already present in nature. To their surprise, they discovered and tagged 233 adult butterflies—ten times the numbers in recent years. Jaret attributed the increase to a wet year, following several very dry years, which promoted the growth of the caterpillar's food and thus the caterpillar itself.

Jaret and his group proceeded to supplement the population. In 2014, they released three hundred caterpillars and forty-six adults. Only a fraction of caterpillars survived, as expected under natural conditions. Their release still signified a big boost to the Schaus' Swallowtail population. In 2015, they released

578 more adults to Elliott Key, Key Largo, and Adams Key. Just before this, Jaret and his group confronted another, previously unimaginable dilemma. Were these numbers so high that they could be causing the population to reach or exceed the maximum level that these small islands could support? If so, further growth could happen only if the range of the Schaus' Swallowtail could expand with more habitat conservation.

As it turned out, these concerns were premature. Jaret's group observed 308 Schaus' Swallowtails in 2015, 68 in 2016, 38 in 2017, and 306 in 2018. Though surpassing the four butterflies captured in 2013, these numbers were alarming, as they did not signify a tremendous recovery. Schaus' Swallowtail's conservation appeared to inch forward only to remain still.

I asked Jaret why numbers had not increased to stable, higher levels. He attributed low numbers to poor weather conditions. For example, the low numbers in 2017 and the high numbers in 2018 occurred in conjunction with low and high rainfall in the respective years. This seemed plausible. Yet I worry about unknown threats correlated with global environmental change. Rearing captive butterflies combined with releasing marked butterflies at different life stages provides the unique opportunity to monitor survival of caterpillars and adults and to track their dispersal within and between populations. With improved techniques and careful observation of the fates of this small population, perhaps this time butterfly releases will stabilize populations of this very, very rare butterfly.

THROUGH THE MOSQUITO HAZE, A BUTTERFLY

After all of these releases, I finally had the opportunity to see Schaus' Swallowtails in the wild in May 2018. I joined a team of three experts tasked with monitoring the population on

Elliott Key in South Florida. As I stepped off a National Park Service boat, Erica Henry handed me a heavy-duty, hooded insect-proof shirt made of dense fabric except for the small area of netting over the face. Thousands of mosquitoes surrounded us. Their incessant buzz persisted for our entire ten-mile trek. The shirt was uncomfortably hot, but it worked. The only mosquito bites I remember occurred after I removed my glove to take photos or unzipped my head net to drink water.

Our trail south took us through a forested tunnel (Plate 13, bottom). Dense stands of thirty-foot-tall trees lined the trail. Their trunks were mainly one to four inches in diameter, and they grew so densely that the forest was nearly impassable. The dense covering of trees formed the hammock. Our narrow path took us down a few feet through shallow, flooded depressions and mangrove forests, and then up a few feet onto higher, drier land.

My only goal on this trip was to see an adult Schaus' Swallowtail. In the process, I helped the group count and mark the butterflies. I learned in the first stretch of our excursion how difficult it would be to accomplish the group's aims. We walked for an hour before Keith Curry-Pochy yelled and pointed upward. I saw a black insect flash down the trail three feet above my head. I could not tell if it was a Schaus' Swallowtail or a different big, black butterfly, a Bahamian Swallowtail (*Papilio andraemon*) or a Giant Swallowtail (*Papilio cresphontes*). As it flew overhead, I swung my net. I thought I was on target. Instead, I hit a tree branch. Twenty minutes later, I saw a second Schaus' Swallowtail. I could not tell if it was the one I had already seen. The butterfly turned abruptly into the forest. It then changed its behavior, and its slow and methodical flight indicated to me that it was a female looking to lay eggs. I crashed through the woods, lunged at the butterfly, and missed.

After lunch, we entered what Keith considered a hot spot of Schaus' Swallowtails. Over the next two hours, we observed eighteen more butterflies. Each sighting started a race to catch the butterfly. Erica swung deftly and caught the first one. Keith caught the next one. He removed his glove to mark it. The impressive number of mosquitoes that immediately latched onto his exposed hand still haunts me. I was the only one who did not catch a butterfly that day. In one final, desperate swing overhead, the net pole crashed into my face and bloodied my lip.

This one survey incorporated two methods used in studies of butterflies, simple counts and rigorous marking studies. It impressed on me yet again just how hard it is to estimate population sizes of the rarest butterflies.

THE RAREST BUTTERFLY

I have concluded that the Schaus' Swallowtail is the very rarest butterfly in the world. This holds true in my comparison of numbers of wild population sizes of the rare butterflies. Two lines of evidence support my conclusion that the Schaus' Swallowtail is the rarest. First, its measured numbers are the lowest. Numbers of adult butterflies counted since 2011 have been lower than four hundred individuals. In some years, the number has declined to fewer than one hundred. In the lowest year, it was four butterflies. By this straightforward count, the Schaus' Swallowtail reaches numbers lower than those measured for any other rare butterfly. It is as close as imaginable to extinction.

A second measure, what conservation geneticists call the "effective population size," yields a population size that is even smaller. The effective population size reflects the number of individuals that contribute genes to the next generation. One measure of the effective population size incorporates the number and balance of males and females in the population. A tiny number of female butterflies produced the few hundred Schaus' Swallowtail individuals flying now. In addition to the two females captured in 2013, there is a small possibility that the seven captured caterpillars could have an equal number of mothers. This yields a maximum of nine females. Of the thirty-two butterflies observed that year, we can make a reasonable guess that sixteen were female.

This small number of females yields a small number of genes passed to future generations. In ecology and genetics, a body of evidence tells us that lower genetic diversity reduces population persistence. By this measure, the effective population size of the Schaus' Swallowtail is much smaller than the number counted in the wild. This remains true even after the population grows. These individuals are the descendants of the same few individuals that started in the lab.

I don't want my comments to convey a lack of enthusiasm for captive-breeding programs. Another tenet of conservation genetics is to promote rapid population growth even from a starting point of low genetic diversity. As populations grow, butterfly genetic diversity can increase over time. In my own experience with the St. Francis' Satyr, a population with small founding numbers can do quite well over time. Perhaps the Schaus' Swallowtail and other butterflies are exceptional cases. To find out, more research is needed. However, the theoretical potential of problems caused by low genetic diversity is poor justification to prevent restoration through captive breeding of the Schaus' Swallowtail. The population size is so low that there is no other way to save the subspecies.

The Schaus' Swallowtail is, and has been for its century-long recorded history, teetering on a fence between persistence and extinction. So far, two historical proclamations of extinction, one in the 1920s and another in the 1930s, have been erroneous.

I'd like to remain hopeful that well-established captive breeding efforts, accumulating scientific advances in understanding the Schaus' Swallowtail's biology, and current habitat protection and restoration can stabilize populations. Armed with this knowledge, we can hold out some possibility that it will establish after release in historic parts of its range.

Ultimately, for true restoration Schaus' Swallowtail populations must establish and stabilize in the wild for decades. What is most discouraging about prospects for the Schaus' Swallowtail is that conservation efforts within South Florida will not be enough. Changes in precipitation, hurricane intensity, and sea-level rise are imminent. Now that Schaus' Swallowtail populations are at such low numbers, extinction cannot be far off. Intervention is essential. Much more effort will be needed to see this butterfly through a second century.

PART II
THE FLIGHT PATH FORWARD

CHAPTER 8

THE FINAL FLIGHT OF THE BRITISH LARGE BLUE

I have never seen the British Large Blue. It is extinct. However, I include it in this book because the case of the British Large Blue holds key lessons for butterfly science and for conservation. Scientists studied the British Large Blue for a century longer than most of the rarest butterflies. They revealed key features of natural history, features that are essential for conservation in general, even if the British Large Blue could not be saved.

The Large Blue was first named by Carolus Linnaeus in *Systema Naturae*, the work that launched the field of zoological nomenclature, written in 1758. At that time, all butterflies received as a genus name the Latin word for butterfly, *Papilio*, a name now applied only to swallowtails. Linnaeus named the Large Blue *Papilio arion*, the species name reflecting his frequent use of Greek mythological characters—Arion referred to the immortal horse of the *Iliad*. The scientific name is now *Maculinea arion*. Of about a half dozen subspecies of the Large Blue found in Eurasia, one, the focus of this chapter, was found only in England; it was named *Maculinea arion eutyphron* (Plate 14). Other Large Blue subspecies outside of England are declining, but none of those is among the rarest butterflies in the world.

The British Large Blue was native to southwestern England, from the western border of London to the peninsula between the English and Bristol Channels. Although its historic range is unknown, records exist from nearly one hundred different populations found in places that harbored the butterfly's host plant, Wild Thyme (*Thymus praecox*). Early on, the greatest threats to the butterfly existed near London. Human populations expanded and transformed the landscape with farm fields, forest planting and recovery, quarrying, and urban expansion. These changes acted in synergy to cause the butterfly's decline.

A CENTURY OF DECLINE

Beginning in the mid-nineteenth century, butterfly biologists recognized low and declining numbers of the British Large Blue. The decline was rapid, leaving entomologists dumbfounded. Some grasslands appeared perfectly suitable, as they supported Wild Thyme and flowers. Yet the butterfly was still vanishing. The discrepancy between biological understanding of high-quality habitat and the downward trajectory of butterfly populations remained a mystery for another century. More science was needed.

The situation was already so dire in 1884 that two lepidopterists, Herbert Goss and Herbert Marsden, wrote articles with titles beginning *On the Probable Extinction of [the Large Blue]*.

Combined, they proposed three hypotheses to explain the decline: 1) cool and damp weather prevalent for several years running; 2) increased burning of grasslands; and/or 3) accelerating interest in butterfly collecting. Weather patterns seemed like a good candidate to explain ephemeral declines within the decade, but great uncertainty remained.

The decline of the British Large Blue continued unabated for the next half century. Through that period, weather fluctuated as expected and did not appear to explain the decline. The fallback explanation became overcollecting. When people saw collectors killing England's beautiful, rarest butterfly, they saw tangible evidence of butterfly death. Entomologists continued to predict extinction. In an attempt to reconcile the long-term decline and impending extinction of the Large Blue, W. G. Sheldon wondered, in 1925: "Will [butterfly collectors. . . . be present] this year to complete the funeral? And if they do will there be a single corpse for interment?"

ON HOST PLANTS AND HOST ANTS

The British Large Blue flew where Wild Thyme was abundant. However, the presence of Wild Thyme did not always correlate with the presence of the butterfly, and scientists were at a loss to explain the most basic elements of the subspecies' biology. Entomologists consistently failed to raise British Large Blues through their entire life cycle from egg to adult on Wild Thyme. After feeding on Wild Thyme in early stages in the wild, older caterpillars vanished. There was a missing piece of information about British Large Blue biology that prevented scientists from explaining how the butterfly persisted in the wild.

I first read in astonishment about the time that elapsed before scientists and butterfly collectors learned key details of the British Large Blue's natural history. Perhaps I should not have been, as this long duration of observation and study is a theme that repeats across the rarest butterflies. The lesson is one of patience, as needed science accumulates to further conservation.

In 1906, Frederick Frohawk made a key discovery that spurred the understanding of the British Large Blue's biology. He excavated ant nests and was surprised to find British Large Blue caterpillars. Close association between ants and blue butterflies is not unique. Indeed, it is rather common. Ants protect caterpillars from potential predators. Caterpillars secrete honeydew, a sugar-rich substance that provides a food source for ants. While the association between blue butterfly caterpillars and ants is common, ants raising caterpillars inside their colonies is unusual. Frohawk assumed that adult ants were feeding ant larvae and Large Blue larvae with the same food. He wrote, "It was with no small amount of satisfaction that we then, for the first time, had before us a natural object which had never been seen by anyone before, and had been wrapped in mystery and remained one of the greatest of entomological puzzles."

In 1915, scientists made two more breakthroughs. In May, Thomas Chapman pulled Wild Thyme plants out of soil that harbored the nest of an ant, *Myrmica sabuleti*. By that time, he was unsurprised to find an older British Large Blue caterpillar living below the soil and within the nest. In the process of ripping up the plant, he killed the caterpillar. Upon examination of the contents of the caterpillar's gut, he made the surprising discovery that the caterpillar had been feeding on ant larvae.

In August 1915, Edward Bagwell Purefoy teamed with Frohawk to make another important discovery. Purefoy wrote to Frohawk and reported on a sequence of events that unfolded after a Large Blue caterpillar dropped from the plant to the ground. An ant tended to it, fed on its honeydew, and then carried it off. Purefoy positioned himself to follow the ant to its destination. Just at that moment, in a remarkable feat of bad luck, visitors from London arrived without notice. They distracted him, and he lost track of the caterpillar. The observation was enough to rekindle hope that a decades-long mystery could be solved.

As is often the case in the progression of science, observations motivated experiments. Purefoy teamed with Frohawk to create experimental studies that tested the role of ants in the development of British Large Blue caterpillars. In doing so, they made another surprising discovery. They built small mounds of dirt and planted them with Wild Thyme. They introduced into the mound one ant species as a suspected partner. They were surprised to observe the ants of another species quickly replace the original ants. Then Purefoy and Frohawk built mounds of dirt against removable boards. By removing the board, they could get an up-close view of what was going on inside the ant nests. They placed older caterpillars on the ground. In each case, an ant found the caterpillar, fed from it, and then carried it off. The caterpillar bulged near its head to signal to the ant, and the ant grabbed it to carry it away. Frohawk and Purefoy eventually discovered the destination. In October, the pair pulled the board away from the mound that supported the ant nest. Just a few inches below the ground, they found British Large Blue caterpillars.

Although Purefoy and Frohawk could raise caterpillars, they initially were unable to raise them through to metamorphosis. This required yet another experiment with middle-aged caterpillars. In typical studies, scientists placed Large Blue caterpillars on Wild Thyme. That was not enough. Another consideration was that British Large Blue caterpillars needed to be on thyme that was in close proximity to ants. Purefoy and Frohawk removed caterpillars from plants and placed one or two in half walnut shells, where they isolated and observed them. In the shells they had a contained environment in which they could feed older caterpillars with ant larvae and contain them while they were in diapause and then as chrysalides. The method was ridiculed by other entomologists at first, before it demonstrated the creativity needed, especially on a low budget, to make scientific progress. In 1918, after three seasons of these experiments, the first British Large Blue completed its life cycle. Several decades later, Purefoy reflected: "It was all intensely interesting and I shall always consider that we were very lucky to have a 'finger in the pie.'"

The British Large Blue's biology and its relationship with ants needed to be unraveled further. The caterpillars exhibited behavioral and chemical signals that entrained the ants into their service. At a certain age, the caterpillars left the Wild Thyme host plant and went to live underground, at the outer rim of an ant nest. They occasionally descended into the nest's core to feed. They fed first on the oldest and largest ant larvae, then switched to smaller ants and their eggs. During their ten months underground from one fall to the next summer, caterpillars grew about one hundred times larger. At times they could suspend feeding while the ant population grew. If the caterpillars ate too many ant eggs and larvae, they depleted their food resource and risked starvation. To avoid this, they would move to a new ant nest.

Why would ants permit Large Blues in their nest? Typical associations of species that intentionally live near each other and intentionally interact are mutually beneficial, an interaction called a mutualism. One type of mutualism involves each species providing food to the other. The British Large Blue fits this model of mutualism, except that there is an additional layer whereby the butterfly harms the ant by preying on it.

New research offered some clues as to why this is a mutualism that may make sense. Wild Thyme emits toxic, volatile chemicals that independently attract the ant and the butterfly. The plants endow the butterflies with a defense against predators. For the ant, the plant provides a place free of competing species that cannot tolerate the toxins. This benefit must outweigh the costs of consumption of ant larvae by British Large Blues.

MISPLACED EFFORTS IN CONSERVATION

Since scientists had fretted about the British Large Blue's decline for over a century, they appeared to have assembled all the necessary information for its conservation. The primary source of butterfly decline seemed obvious. People destroyed grassland habitats as they plowed more land and expanded towns and cities. Even where the impact of people was not so direct, habitat degraded as forests grew on former grasslands. Shrubs and trees grew to exclude Wild Thyme, ants, and other features of open habitats that British Large Blues required. It was clear to everyone that protecting native habitats from new development should lead to conservation of the British Large Blue.

It did not. Even in protected grasslands populated by Wild Thyme and ants, butterfly populations continued to decline. Of the ninety-one populations known in England, over half were lost by the 1950s in places where scientists had prescribed methods for land preservation.

The British Large Blue is emblematic of a common mistake made in the conservation of rare butterflies. It is easy to cast blame on the gravest threat, habitat loss. Once that is addressed, butterfly conservationists then look toward collectors that cause immediate harm or harmful, widespread insecticides. In 1980, Jeremy Thomas observed that "most of the measures taken [in the 1960s] can now be shown to have been irrelevant or actually harmful to the butterfly's needs." The hypothesis that habitat loss caused Large Blue population decline remained to be tested. Still more science was needed to guide conservation.

As the butterfly declined to dangerously low levels in the 1960s and 1970s, research intensified. Jeremy Thomas and others invested great efforts to map out the precise life cycle of the British Large Blue. Although they had identified details of the mutualistic relationship between the butterflies and ants, they had assumed that different ant species were involved in the interaction. When habitat protection commenced, the focus was to maintain this set of ant species.

Over time, scientists narrowed the possible set of suitors to two different red ant species in the genus *Myrmica*. Neither of these ants was rare. Yet, not all ants are created equal. Through careful investigation, especially beginning in the 1970s, scientists found that one species of red ant, *Myrmica sabuleti*, tended caterpillars and provided them food. Any conservation for the British Large Blue depended on the presence of not just any ant, even if closely related, but this particular ant.

This new understanding refocused conservation efforts. In grasslands replete with Wild Thyme, other factors affected the presence and abundance of this one ant species. The ant preferred physical environments on steep, south-facing slopes with warmer climates.

In their initial efforts, conservationists thought a grassland was a grassland. I understand this. When I travel to seek rare butterflies, I use visual habitat cues to hone my search. As a naturalist, I feel confident in my ability to discern high-quality habitats for any species of butterfly. As it turns out, I am not as good as I think I am. Often, very subtle differences in structurally similar habitats tilt the balance between suitable and unsuitable areas. The case of the British Large Blue has caused me to reassess my ability to restore habitat for other species that have been my focus for two decades.

What emerged as a key quality of habitat for the Large Blue and for *Myrmica sabuleti* was the height of grasses and other vegetation. The grasslands changed in subtle ways over time. Disturbances created, destroyed, or modified grasslands, including grass height. After fire or clearing, a succession of plants proceeded over time from small grasses to large grasses to herbs to shrubs and then to trees. *M. sabuleti* occurred in only a narrow—almost imperceptible—band of this succession when grasses were less than 2.1 inches high. As grasslands grew, other ant species replaced *M. sabuleti*. Disturbance was required to maintain grasses of low stature. The implication was that habitat replete with Wild Thyme and ants in general was, from the perspective of the British Large Blue, not quality habitat.

The failure to recognize the subtleties of disturbance needed to maintain populations allowed the British Large Blue's decline. The one natural force in England that could have maintained low vegetation was herbivory. An important herbivore in this ecosystem was the European Rabbit (*Oryctolagus cuniculus*), a species introduced from the continent hundreds of years ago. The rabbit and the Large Blue declined simultaneously. A once common species, the European Rabbit succumbed to a deadly viral pox, myxomatosis, which arrived in England in 1953. The disease had spread in the 1890s from South America to other regions of the world, including continental Europe. People established it in France to control rabbits. Its arrival in England was imminent. Once there, people deliberately spread the disease to reduce rabbit populations to a tiny fraction of their historic size.

Even before the virus became endemic, the rabbit was already a less effective grazer of British Large Blue habitats. Ignorant of the connection between butterflies, low-stature grasslands, and ants, conservationists restricted access of animals perceived as competing with the butterfly for its host plant. This included herbivory by the rabbit and, as its populations plummeted, by grazing cows. These cows had the potential to cause even more damage to the ecosystem because of trampling. The solution: fence in British Large Blue habitat. This excluded both herbivores and another perceived threat, butterfly collectors. Paradoxically, conservationists intentionally excluded disturbance at the time it was needed most. In doing so, they caused butterfly habitats to degrade and butterfly populations to be lost.

The British Large Blue entered the extinction vortex, whereby environmental and biological changes reinforced one another and caused small populations to become smaller still. By the early 1950s, the British Large Blue inhabited thirty grasslands. By the mid-1960s, the number of populations had dwindled to four. With a regular cadence, butterfly populations were lost in 1967, 1971, and 1973.

Frustratingly, scientists learned key details of British Large Blue ecology when just one small population remained. By the time the science was known, it could not address the problem of a population that was so small. The scientists managed the last population's habitat, and numbers increased. However, the population rose to a level that the ants could not support. To understand this dynamic, scientists created artificial environments in a lab with British Large Blue larvae and ant colonies. They found that butterfly numbers rose so high that the butterflies extinguished the ant colony and themselves. Severe droughts in the mid-1970s then reduced the butterfly population to handfuls of individuals. Once this small population entered the extinction vortex, random events befell it. One of these events occurred in 1977. Of fifteen remaining individuals, there were only three females, thus limiting egg production. In 1978, the population numbered only five adults (of which two were female) that bred in captivity. These generated twenty-two adults. In my experience, it is difficult to coax small numbers of rare butterflies held in captivity to mate. Thus, I was not surprised to learn of the same problem with these few British Large Blues. It precipitated the final flight of the British Large Blue. In 1979, it was extinct.

Subspecies *Maculinea arion eutyphron* no longer exists on the earth. Conservation of this butterfly was on British entomologists' and naturalists' minds for over a century. Despite the measures taken to protect the butterfly, they did not know enough to save it.

CONSERVATION BY WAY OF SUBSPECIES

How then can it be that after this tragic history, the Large Blue flies today in England? These are not the *British* Large Blues; they are a different subspecies. Before extinction, the science had advanced to the point that a new and viable conservation path was possible in England. Restoration could create the right habitat with the right plants and the right ants to sustain Large Blues. There was just one thing missing—the butterfly.

As I mentioned at the beginning of the chapter, the British Large Blue was one of about a half dozen subspecies of Large Blues. Scientists and conservationists devised a plan to capture individuals of the Swedish subspecies, *Maculinea arion arion*, and introduce them to newly restored sites in England. Beginning in 1983, they launched efforts at three sites. Over the next two decades, they seeded populations at even more sites. Butterflies established in many of them. By 2008, the Swedish subspecies of Large Blue had established populations in twenty-five sites, a number similar to the populations of British Large Blue found in 1950. The size of the largest population exceeded one thousand butterflies, the largest size ever observed.

Was the Large Blue fragile or resilient in the face of environmental change? The answer is mixed. Its fragility was most apparent, evident in its extinction. The British Large Blue was unable to withstand persistent change to its habitat, including degradation of the community of plants and animals on which it depended. For over a century, expanding human populations whittled away its habitat by intensifying land use. Changes were intentional but misconceived (fencing out people and at the same time needed grazers) or unintentional (spread of myxomatosis to kill rabbits, not butterflies). Fragility of the British Large Blue prevailed.

At the same time, Large Blues have proved resilient. Another subspecies, the Swedish Large Blue, resuscitated the Large Blue's presence in England. People eventually possessed the science to inform them of the species' biology and needs. This translated to the right suite of tools for conservation. The solution was not as simple as protecting habitat. The key ingredients included a community of interacting plants and animals.

Even though it came too late, aspects of the recovery effort were a success. Both the extinction of the British Large Blue and recovery of the Swedish subspecies of Large Blue hold key lessons for those seeking to conserve the rarest butterflies in the world. There is one lesson that the British Large Blue is shouting: it takes tremendous investment of effort and time to understand how to conserve and restore populations of the rarest butterflies. Over a century of concerted effort was devoted to the British Large Blue. The story of the British Large Blue, its life history, and its behavior are complex. Surprises about the butterfly's biology emerged regularly for over a century, yet they were not enough.

Given the idiosyncratic features of this subspecies' biology and behavior, I wonder what aspects of the lives of other rare butterflies may be eluding our notice. Although my focus is on the rarest butterflies, I suspect that a poor understanding of natural history is likely a common feature in our struggles with conservation of many of the rarest insects and plants. I perhaps incorrectly imagined the declines of the rarest butterflies would be akin to declines of other extinct species whose losses appeared simpler to explain. For example, many birds, mammals, and plants on remote islands have disappeared quickly in the last half millennium, caused in many cases by novel predators. The history of the British Large Blue demonstrates that the path to extinction is not simple and that recovery requires detailed knowledge of natural history and global change.

I have dedicated two decades to the study of the St. Francis' Satyr, the Crystal Skipper, the Miami Blue, and the Bartram's Scrub-Hairstreak in the hope of securing their existence. I would like to think that, for these butterflies, the path to conservation follows the one laid by subspecies such as the Fender's Blue, for which key aspects of natural history, discovered within a couple of decades, set populations on an upward trend. If the butterflies I study follow the story of the British Large Blue, I may not live long enough to see their recovery.

I have ended each chapter with the question: Is this butterfly the rarest species in the world? The British Large Blue held this position in the late 1970s, before it went extinct. Hopefully, the

Schaus' Swallowtail and the other rarest butterflies will not share the British Large Blue's fate.

This gives rise to a final lesson I draw from the British Large Blue: efforts to stop species extinction and promote recovery cannot start too soon. Earlier and focused science might prevent the rarest butterflies from descending to extinction. Understanding the ecology of rare butterflies may help their recovery. We need to remember this lesson for the conservation of very common butterflies whose populations are declining.

CHAPTER 9

MONARCHS

THE PERILS FOR ABUNDANT BUTTERFLIES

The Monarch is one of the two largest butterflies in North America, vivid in color and graceful in form (Plate 15). Its broad distribution and the great migration of the Eastern North American population make it the best-known butterfly and one of the best-known insects in the world. Unlike the rest of the butterflies in this book, which number in the hundreds or thousands, Monarchs number tens to hundreds of millions. The Monarch stands in stark contrast to the rarest butterflies.

The Monarch (*Danaus plexippus*) has an extremely large geographic range, larger than the ranges of nearly all the world's butterflies. It occurs through most of the Western Hemisphere, in North America from Canada and the United States to Mexico and then south throughout Central America and into South America north of the Amazon River. The Monarch's distribution extends to Hawaii, Australia, New Zealand, and various Pacific Islands, and across the Atlantic Ocean to Portugal and Spain and occasionally other European nations.

Compared to other butterflies, including Monarchs in other regions, the Eastern North American Monarch population stands out because of its great migration. Each year, it travels from northern North America to central Mexico and back again. In the fall, the trip south is up to three thousand miles long. In the spring and summer, the same distance is traveled but in successive generations. An important point is that the individuals returning to Mexico are several generations removed from those that departed. For all these reasons, the Monarch migration is distinctive.

The migration involves two areas: a small overwintering range in Mexico and a broad breeding range in eastern North America north of Mexico. The winter range is the destination of tens of millions to hundreds of millions of Monarchs. It is in a small area of high-elevation Oyamel Fir (*Abies religiosa*) forest in the Trans-Mexican Volcanic Belt, a mountainous region in central Mexico. There the butterflies occupy forest areas separated by short distances; the butterflies clustered in each area are referred to as a colony.

Monarchs land on and suspend themselves from large Oyamel Fir trees that, when mature, approach one hundred feet in height. These trees occur in mountainous terrain where peaks rise to nearly two miles above sea level. Butterflies congregate on steep, southwest-facing slopes. Monarchs cannot tolerate the freezing temperatures found in more northerly North American winters, motivating their migration to milder climates. The overwintering destination experiences temperatures close to, but still above, freezing (except in rare circumstances) and has a suitable climate to protect the Monarchs through the winter. The Monarchs arrive in Mexico in October and pass the winter in dense clusters on trees.

Monarchs complete the rest of their life cycle on more northerly breeding grounds. The generation that departs from the Mexican overwintering sites travels to the southern United States to lay eggs and complete one generation. The following generations sweep northward, arriving as far north as southern Canada. The overwintering Monarchs depart for the north by the end of March. These Monarchs and the following generations time their arrival in their breeding grounds to the place and time that their host plants, milkweeds (*Asclepias* species), are growing.

IS THE MONARCH RARE?

Despite its differences from the rarest butterflies, the Monarch of eastern North America may eventually be threatened with extinction. The number of Monarchs in eastern North America has declined precipitously. The population could foreseeably join the list of the rarest butterflies. Because the entire population congregates in the same locations in winter, the number of Monarchs overwintering in Mexico provides a standard to assess population trends. (For simplicity, I will refer to the Eastern North American Monarch as the Monarch.) Each year, scientists survey the area occupied by Monarchs. Scientists have estimated that fifty million individuals blanket each hectare (two and a half acres). These butterfly clusters spread across many forest groves. The area covered by this mass of Monarchs provides an index of the number of Monarchs that arrived safely following their long migration.

These data paint a bleak picture. There are no data from before the discovery of the overwintering grounds in 1975.

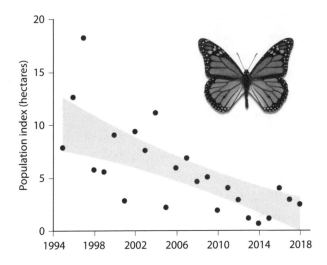

FIGURE 9.1. Index of population size of Monarchs in their overwintering range in Mexico. Scientists have measured the number of Monarchs per hectare (two and a half acres) to equal fifty million. Thus, ten hectares (twenty-five acres) support five hundred million Monarchs.

Historical anecdotes suggest high abundances, including nineteenth-century reports that Monarchs were so numerous they obscured the sun during their migration. At the highest point of quantitative measures, in the winter of 1996–97, Monarchs covered about eighteen hectares (almost forty-five acres) with about a billion butterflies. Since then, populations have plummeted (Figure 9.1). In the winter of 2013–14, scientists reported the lowest overwintering area (two acres) and population size (thirty million butterflies). The population size had dropped by more than 97 percent.

Since then the numbers have rebounded to some extent. They rose to 200 million butterflies in 2015–16 and then declined to 125 million in 2017–18. These large fluctuations are characteristic of butterfly populations. Female Monarchs lay many eggs. Because there are three to five generations per summer, a string of productive generations within one year can boost population sizes quickly. Just as quickly, numbers can plummet. Lincoln Brower and his colleagues at Sweet Briar College, Virginia, reported a large, natural decline over the winter in 2002. On January 12, a storm front dropped two inches of rain on the roosting Monarchs. The rain then turned to snow, and another two inches fell. Afterward, dead butterflies covered the ground to depths of nearly two feet. Brower estimated that the storm killed about half a billion Monarchs. The Monarchs recovered the following year. Still, the event showed how rapidly a great number of butterflies could be lost.

Although fluctuations are expected, they present a major danger, especially as populations get smaller. Over time, greater variation in population size and population growth rate has the effect of reducing the population growth rate. We know this from the first principles of statistics. Fluctuations upward can rise to high levels; fluctuations downward cannot go below zero (extinction). The consequence is that the overall growth rate is lower than the mean value, and this pushes population sizes lower and lower. Imagine a storm like the one in 2002; one with a toll of similar magnitude now could wipe out the entire population. Taken together, natural variation in and growing threats to Monarchs have caused declining trends over the past two decades.

Although the focus of this chapter is on the Monarch population of eastern North America, the Monarch population of western North America has shown similar trends. By current accounts, the western population is in even greater jeopardy than the eastern population. Cheryl Schultz and colleagues estimated that, since 1986, its numbers have fallen to less than 5 percent of historic numbers (dropping from about 4.5 million to 200,000 individuals). At this rate of decline, they estimated that the Monarch of western North America has a 72 percent chance of extirpation within twenty years and an 86 percent chance of extirpation within fifty years.

In the book's introduction, I intentionally distanced myself from Monarchs. The population size of the Eastern North American Monarch is so big, its summer range is so immense, and its biology is so different that comparing it to the other rare butterflies is like comparing apples to oranges. Monarchs in eastern North America outnumber any of the rarest butterflies by at least ten thousand times.

Even with these vast differences between Monarchs and the rarest butterflies, I now recognize striking similarities. Although the areas of their summer ranges are incomparable, the area of the winter range of the Eastern North American Monarch is fifty acres, on par with the ranges of some of the rarest butterflies, such as the St. Francis' Satyr and the Miami Blue. The Monarch is as vulnerable to the loss of Mexican forest as other butterflies are to the loss of small areas of their habitats.

Another similarity is in population trajectories. The rarest butterflies today have populations that number a tiny fraction of their former size. To get there, their population declines must have exceeded the 87 percent or greater loss we have seen in Eastern North American Monarchs and the 97 percent decline in Western North American Monarchs. After twenty years of decline, there remain some one hundred million Monarchs. Consider another 90 percent drop, and then another, and then another. Under this gloomy scenario, the Monarch population of eastern North America would join the ranks of rare butterflies. Like the rarest butterflies and the Monarch of western North America, they would be threatened with extinction.

ADDRESSING THREATS

People are transforming landscapes in both the vast areas of eastern North America where Monarchs breed and in the tiny areas of forest in Mexico where they overwinter. This has been true in North America for centuries. In the past few decades, the effects have been magnified with habitat loss caused by increased human population growth, food consumption, and agricultural intensity. The decline of the Eastern North American Monarch can be explained only by understanding the vastly different threats faced in the overwintering range and the breeding range.

Although the Monarch population size has been falling, scientists have been unable to pinpoint exactly why. The Monarch is enduring at least three upheavals. The decline in Monarch abundance has corresponded with the use of genetically modified, "Roundup-ready" soybeans, introduced into the US in 1996, that are tolerant of herbicides (including the brand Roundup) that eradicate non-crop plants from fields. Because of agriculture's escalating efficiency, there is not much room left in or around those fields for Monarchs and their milkweed host plants. Coincidentally, along with the deployment of Roundup-ready crops, neonicotinoids became the world's most common insecticide. This class of insecticide is slow to break down and, because it is water soluble, flows through plant roots and into their leaves. Experiments with Monarchs have found that tiny doses of neonicotinoids can decrease caterpillar survival. The rise in use of herbicides and insecticides is correlated with the decline in Monarchs (and with the phenomenal decline in abundance of rare butterfly species that live adjacent to the northern arc of the corn-growing regions in the Midwestern United States, including the Dakota Skipper, *Hesperia dacotae*, and the Poweshiek Skipperling, *Oarisma poweshiek*). The total effect of these chemicals on Monarchs is likely to be known only after it is too late.

Second, there have been multiple waves of landscape transformation across the Monarch's breeding range. As people converted forests to farmfields in the northeastern United States through the nineteenth century, they created openings that were likely beneficial to grassland plants such as milkweed. As people spread westward in large numbers in the nineteenth century, they converted prairie that harbored milkweed to farm fields. In the past century, the northeastern United States has experienced increased urbanization and increased forest cover following agricultural abandonment. These changes causing increased and decreased Monarch habitat have had an unknown cumulative effect.

A third threat is to the forests where Monarchs overwinter. The Monarch Butterfly Biosphere Reserve encompasses over one hundred thousand acres. Of that area, thirty thousand acres make up the core zone for Monarchs. Within the core region, more than five thousand acres, or about one-sixth of the total area, have been degraded or lost. A small area of forest loss will cause a disproportionately high loss of the Eastern North American Monarch population.

Scientists debate where along the Monarch's three-thousand-mile journey its population is most threatened. Pinpointing the location and the mechanism of decline has proved elusive. One limit to our knowledge is the restricted area over which research focuses. Scientists measure the total size of the Monarch population at one place on its journey, Mexican forests. What remains to be learned is how loss in Mexico translates to the performance of Monarchs when and where they lay eggs, feed, metamorphose, and migrate. Because of the population size and range of the Monarch, this area of study is daunting.

CITIZEN SCIENCE

In lieu of collecting data across the *entire* Monarch breeding population, scientists have relied on important sources of data from impressive programs in citizen science. Through these programs, a large number of nonscientists collect data more extensively, throughout the Monarch's range. Monarchs have proved to be excellent ambassadors for getting people involved in data collection. Many Monarch scientists, especially University of Kansas professor Chip Taylor and University of Wisconsin professor Karen Oberhauser, have created citizen-science programs, such as Monarch Watch, to track migrating adults, and the Monarch Larva Monitoring Project, to track survival of caterpillars. Such programs provide structured and accessible ways for thousands of people to get involved throughout the range of the Eastern North American Monarch.

In another citizen-science effort, the North American Butterfly Association (NABA) coordinates an annual fourth of July butterfly count, in which volunteers count butterflies in approximately five hundred sites spread across North America, many of which are in the range of the Eastern North American Monarch. On one day each year at each site, volunteers search within a fifteen-mile-diameter circle and count all the butterflies they see. The program is more than four decades old, and its data provide a helpful, coarse-scale barometer of change.

Other, more intensive sampling projects are possible within smaller regions. One exemplary effort within the Monarch's range occurs in Ohio. For two decades, the Ohio Lepidopterist Society has organized volunteers to count all species of butterflies weekly from spring until fall along set routes across the state. One use of these data is to test for effects of climate change on Monarch populations. For example, Michigan State University professor Elise Zipkin used these data to find that warmer temperatures could increase Monarch population growth but that very high temperatures, which are increasingly common, can reduce Monarch population growth, as they are lethal to development. Similar statewide efforts occur within the Monarch range, concentrated in the midwestern United States.

Citizen science efforts are powerful resources for understanding population status, dynamics, and threats throughout the breeding range of the Monarch. Ecologists have begun to assemble these disparate resources to explain the locations and times of threats that reduce Monarch populations. Georgetown University professor Leslie Ries led one team that found positive correlations among measures of population size taken throughout the Monarch's path, from overwintering grounds to northern breeding grounds. They discovered that the relationship fell apart on the return trip to Mexico and hypothesized that the butterflies were experiencing low survivorship during migration southward. Sarah Saunders of the National Audubon Society led a team that found that population sizes at the overwintering site in Mexico are affected by 1) the summer population size in the breeding range; 2) the availability of nectar resources during fall migration; and 3) the amount of forest habitat available in the overwintering range. These types of analyses have directed attention to new possible threats.

Such studies may identify pathways to conservation that stabilize the Monarch population. The root causes of decline may need additional layers of experimental or observational studies at very small scales. I share the Monarch Joint Venture's aspirational view. The Monarch vies with honeybees and just a few other species for the world's best-known insect. Standardized protocols for regular, rigorous data collection across the summer range would permit estimates of range-wide population trends and would reveal regional areas where the Monarch may experience different threats. These efforts could piggyback on surveys like those in Ohio, surveys that seem likely to expand to other states.

A THREATENED SPECIES, A THREATENED MIGRATION

The Eastern North American Monarch's plummeting population propelled a team of butterfly conservationists to demand support for broader action. In August 2014, the Center for Biological Diversity, the Xerces Society, and the Center for Food Safety, as well as the Monarch's champion scientist, Lincoln Brower, submitted a petition to the US Fish and Wildlife Service to request listing *Danaus plexippus plexippus*, the subspecies of the Monarch that migrates in eastern and western North America, as a threatened species. On December 31, 2014, the US Fish and Wildlife Service agreed to a status review that would include a more detailed evaluation and comments from any interested party. This is typically a ninety-day process, but the US Fish and Wildlife Service was granted a legally binding extension to gather more information; it will make a decision in 2019.

Should Monarchs be listed as threatened, which implies they are vulnerable to becoming endangered in the near future? If population size is the criteria, then the Monarch will not be listed. The US Endangered Species Act states no criterion for population size for any plant or animal species that ensures recognition. Other animals listed as endangered occur at much higher population sizes than butterflies on the list. As one example, endangered salmon can number into the hundreds of thousands. At numbers exceeding thirty million in eastern North America, the Monarch is at least ten thousand times more abundant than other butterflies recognized on this list. Other butterflies that are much rarer than Monarchs receive no protection.

There are other quantitative and more defensible criteria for protection. The Red List of Threatened Species maintained by the International Union for Conservation of Nature (IUCN) has many quantitative criteria relative to population size and range size. For example, any species whose global numbers have declined 90 percent or more achieves the status of critically endangered. When threats and solutions are unclear, species achieve the same status after an 80 percent decline. The Eastern North American Monarch population has declined by 87 percent or more; the Monarch population in western North America has declined by 97 percent. Perhaps the US Fish and Wildlife Service will consider these defensible, quantitative criteria for listing the Monarch.

While the process to list the Monarch moved forward in 2015, the White House launched a new initiative to promote Monarch recovery. One key focus of this initiative was conservation of a Monarch "superhighway," or corridor, that connects habitats throughout its migration route. The concept was to restore habitat in the area around the interstate (I-35) running from Minnesota to Texas. The idea of landscape corridors that connect otherwise fragmented habitats is not new. Indeed, it is one focus of my ecological research. Corridors often follow natural features such as rivers or mountain ridges or take advantage of greenways. What was unique about this vision was its sheer length, crossing the entire north–south extent of the United States.

The White House's task force did not focus on a standard, scientific definition of a corridor. Corridors fitting a scientific definition provide a true path that gets an animal or plant safely along the entire distance from an origin to a destination. With respect to the rarest butterflies, my lab has shown that forests along streams provide corridors for St. Francis' Satyrs to move between grassy wetlands. The effort for Monarchs does not fully connect the path between most breeding grounds (which cover a massive area) and the overwintering home in Mexico. This initiative focused on the land within a mile to either side of the interstate. This path has benefits. It follows at least roughly along a key part of the route of Monarch migration across much of its distance and counteracts the effects of milkweed loss there.

Of course, highways have one major flaw. Interstates support a steady flow of traffic that threatens survival of migrating Monarchs. In one scientific study that evaluated the effects of roadside grasslands for butterflies in Iowa, Leslie Ries found that butterflies move along roads but rarely across them. She found that the benefits of these plantings for butterflies outweigh their costs. In a massive undertaking like the Monarch corridor, new studies will determine risks.

Even among the most enthusiastic lepidopterists and conservation biologists, there is not uniform support for listing the Eastern North American Monarch as threatened. This has surprised me. I have asked many ecologists and restoration biologists active in butterfly and Monarch conservation for their views. The issues are complicated. When I asked Jaret Daniels, he posed questions such as: Would children still be able to learn about a Monarch by catching it or raising its caterpillars? Would milkweeds and Monarchs in someone's yard bring new limitations on land use? In these cases, exemptions from regulation are likely, but they would likely cause unnecessary confusion. Are regulations associated with listing even feasible, given the extent of farmland in the Monarch's range? When I asked Karen Oberhauser for her views on listing the Monarch, she told me how she had wrestled with these issues. Ultimately, she feels that given the current status of the population and precipitous declines over the past two decades, the Eastern North American Monarch warrants listing under the Endagered Species Act.

As I said, Monarch population numbers are nowhere near those of the rarest butterflies in the world. Given this, it seems absurd to imagine them becoming rare. Yet, there are prominent examples that suggest it is possible.

Every year in my introductory undergraduate class in ecology, I describe how extinction is possible even for the most abundant species on earth. The iconic example is the Passenger Pigeon (*Ectopistes migratorius*). A century and a half ago, it was the most numerous bird in North America. During Passenger Pigeon migrations, massive flocks filled the skies. Like the Monarch, the birds numbered in the hundreds of millions. In the latter half of the nineteenth century, trains and telegraphs accelerated human settlement, forest loss, and overhunting. Passenger Pigeon populations collapsed in just a few decades. By 1914, the last individual, known as Martha, died in the Cincinnati Zoo, and the species was extinct. Are Monarchs the next Passenger Pigeons?

The time to address declines in the Monarch population is now. As University of Waterloo (Canada) professor Thomas Homer-Dixon wrote, "Once a system shifts to a new state—once the Monarch migration collapses, for instance—it's usually very difficult, even impossible, to tip the system back to its earlier state. The system fundamentally reconfigures itself, and the new arrangement can be very stable." The Monarch fits within this book, not as a counter-example or as one of the rarest butterflies in the world. Rather, it provides a call to action to prevent the list of rarest butterflies from expanding. Monarchs are torchbearers for countless other butterflies (not to mention other insects, vertebrate animals, and plants) that struggle to maintain healthy and stable populations.

CHAPTER 10

THE LAST BUTTERFLY?

I began writing this book with the goal of answering one question: What is the rarest butterfly in the world? I found my answer: the Schaus' Swallowtail. Unfortunately, the list of butterflies that could compete for this title is expanding. This stems in part from a progressive increase in knowledge about butterflies from well-known to lesser-known species or subspecies, a progression that can be represented by their visual appeal. The rarest butterflies discovered a century ago were large and showy. In the middle part of the century, new species of rare butterflies were smaller (but still colorful). In recent years, new species of rare butterflies have been small, brown, and nondescript. If the past century serves as a barometer, the outlook for discovery of more rare butterflies is uncomfortably strong. As global change accelerates, we can anticipate the list of rare butterflies to swell.

When viewed over a century, the trajectory of the rarest butterflies can be summarized in only one direction: downward.

Repeatedly, and for a host of reasons, populations have been lost. With each successive loss, the area occupied by the rarest butterflies became smaller. Now for each butterfly species or subspecies, the number of individuals, the number of populations, and the extent of the area they occupy have become in most cases minuscule. The rarest butterflies are fading from view.

Each year, as I begin sampling the rarest butterflies I study, I walk into their habitat wondering: Have I already seen the last butterfly? In my career, I have not witnessed the extinction of a butterfly species or a subspecies. I have, however, seen the loss of entire populations. When I started my research on rare butterflies, I never imagined I would observe so many populations lost. In twenty years, I have observed the loss of three of seven St. Francis' Satyr populations, one of three Miami Blue populations, and one of a handful of Bartram's Scrub-Hairstreak populations. For the Bay Checkerspot and—especially—the Schaus' Swallowtail, one viable population remains; loss of it would equate to extinction.

I began writing this book with a focus on the very rarest butterflies, those most at risk of extinction. In the short time since I started writing, there have been new data about severe declines of other butterfly species. Reports emerged that the population of the Eastern North American Monarch had dropped in size to approximately two hundred million, then one hundred million, then thirty million, before bouncing back to one hundred million. More generally, a study showed that average population sizes of all butterfly and moth species globally had dropped by 30 percent over forty years, and another showed a 75 percent loss of insect biomass in protected areas in Germany over twenty-seven years.

The last butterfly, in the general sense, is at the same time preposterous and real. It is preposterous because there will be butterflies of some types flying into any imaginable future. Some of these butterflies are better adapted to an earth that people have modified. For example, I can't imagine the last Small Cabbage White, a butterfly species native to Europe, Asia, and North Africa, which is now present in North America and Australia. It was introduced along with its many host plants, which include cabbage, broccoli, and cauliflower.

However, for a growing set of species, even common species, the last butterfly is a real possibility. I'd like to imagine that the Monarch is a unique case of a common butterfly that is in rapid decline. Perhaps its migration to an overwintering range of limited acreage aligns its threats with those of the rarest butterflies. It is doubtful that it experiences threats across its breeding range that are unique to it. Species that were common and widespread in the Monarch's breeding range are now on the verge of extinction. Take the Dakota Skipper and the Poweshiek Skipperling (Plate 16). Just twenty years ago, they thrived in hundreds of populations spread from Michigan to Manitoba. Now they persist in just a handful of scattered populations. As with the Monarch, the cause of their sudden decline remains a mystery. For both, the last butterfly may not be so far off. How many other butterflies and, for that matter, insects are on this same trajectory?

THE NEW CANARIES

The rarest butterflies are, to use a cliché, the canaries in the coal mine for insects, by far the most diverse species group (excluding microbes). As scientists have learned, the rarest butterflies share causes of decline with other insects. Butterfly declines and extinctions are near-certain indicators of other, unobserved extinctions in all other insects.

Compared to what we know about butterflies, we know virtually nothing about the reasons for population change in the other insects. The diversity and, in many cases, abundances of butterflies have been known and tracked for centuries. There are growing numbers of studies across the world that track butterfly population change over time.

As an example of just how little is known about the threats to other insects, I reviewed the scientific literature for all insects that are ranked as critically endangered or endangered by the most reputed conservation organization, the IUCN. I found data on detailed biology of only 6 of 730 species, and even these data were scant.

Just as it is for any other threatened plant or animal, the single greatest threat to butterflies is loss of habitat. Butterflies that reach the list of the rarest species have seen their habitats converted to cities, highways, or fields. Hardwood hammocks of the Schaus' Swallowtail and pine rockland of the Bartram's Scrub-Hairstreak were cleared to create Miami. A highway bisected a population of Bay Checkerspots. Most tragically, San Francisco grew over the top of the habitats of the Xerces Blue (see chapter 1) and the Sthenele Satyr (*Cercyonis sthenele sthenele*) and caused their extinction.

For the rarest butterflies, habitat loss does not exist in isolation. Threats act in concert and actually magnify other negative effects on populations to create a downward spiral. For the Bay Checkerspot, for example, the highway that divided a population brought cars, and cars brought nitrogen pollution, which in turn fertilized serpentine grasslands and shifted the balance of the ecosystem from native to exotic grasses and forbs. In the coming years, climate change will affect butterflies in a variety of ways. Indeed, it has already altered patterns of rainfall and temperature to the detriment of the Bay Checkerspot. These threats become worse when set against a background of habitat loss and fragmentation. Undoubtedly, other insects and plants decline in response to the same threats, and butterflies serve as a broader indicator of biodiversity decline.

DISTURBANCE IS THE THREAD THAT BINDS THEM ALL

As I researched the rarest butterflies, the single biggest surprise was one consistent source of degradation of habitat: loss of natural disturbance. There are three corollaries to this. First, this phenomenon is general, in that all of the rarest butterflies require natural disturbances to maintain their habitat (Figure 10.1). Second, the source of disturbance could differ vastly between one rare butterfly and the next. Third, disturbances are at the same time lethal and essential for stable populations of rare butterflies. Rare insects are likely to respond to disturbance in the same way as butterflies, making butterflies a strong indicator of a disturbance's effects on the broader insect community.

    I grappled with my own bias against disturbance as I researched the rare butterflies. Earlier, I never condoned apparent destruction of butterfly habitat, and by association rare butterflies, and I actively opposed it for more than a decade. However, my thinking transformed as I watched the loss of one population after another. I came to realize that I should not consider conservation of each individual butterfly. I refocused on entire butterfly populations and on persistence of species. Disturbance that caused immediate harm to some individuals was the only way to preserve whole populations; a small part of the population was sacrificed for the greater good. Persistent decline forced me to adopt the unconventional and counterintuitive logic that killing some butterflies can save butterflies.

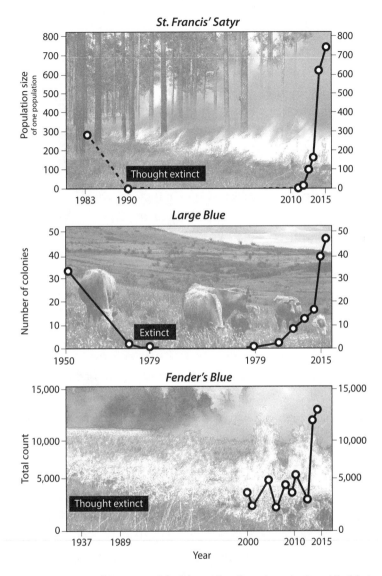

FIGURE 10.1. Population sizes of three butterfly subspecies over time. All of their populations require disturbance, which land managers eventually applied to the landscape. Before disturbance management: The St. Francis' Satyr was lost. The British Large Blue went extinct. The Fender's Blue existed at low population sizes. For the St. Francis' Satyr and the Large Blue, recovery of populations after disturbance management was followed by introduction of individuals from elsewhere.

Here I need to head off a misperception that could precipitate disaster. Taken to extremes, simultaneous disturbance in all of the small areas where the rarest butterflies live would also cause extinction. There is a subtle twist to restoration for the rarest butterflies. Some—but not all—areas need to be burned, flooded, or otherwise disturbed in rotation. For the Fender's Blue, this means one-third of the habitat every year. For the St. Francis' Satyr, this might mean one-fifth of the habitat every five years. The frequency and intensity varies, but the principle is the same. People have tended to reduce the destructive power of natural disturbance that becomes a nuisance or even harmful to property and people. The consequence is that even actions intended for conservation have reduced the quality of habitats for the rarest butterflies. Conservation for the rarest butterflies needs to occur in interconnected landscapes—in metapopulations—that encompass habitats that are and are not disturbed.

Disturbances can be forceful or subtle. Examples include frequent fire to maintain habitats for populations of the Fender's Blue, hurricanes or other strong storms to rejuvenate hardwood hammocks for the Schaus' Swallowtail, and herbivory to maintain populations of the Large Blue. At times, forces of disturbance act in concert. The St. Francis' Satyr lives in wetlands where herbaceous plants regenerate after beavers create dams; in dry years, when beavers are elsewhere on the landscape, fires can also eliminate dense trees and shrubs. Where invasive grasses overrun meadows inhabited by Bay Checkerspots, herbivory can exclude unwanted plants. In nearly every instance, the rarest butterflies benefit from natural disturbance.

In general, the beneficial effects of natural disturbance for the rarest butterflies lead to a clear message for conservation and restoration: take off the kid gloves. The rarest butterflies live within ecosystems that are dynamic, where natural succession causes habitat to become progressively less suitable. Disturbance resets this process. To an ecologist, this all makes sense—it is grounded in basic principles. In the places where rare butterflies live, however, these principles often decouple from conservation action. Too often, attempts to restore disturbance occur at locations distant from where rare butterflies live, or outside the range of their dispersal. Where natural disturbance is not possible, restoration action must replace it.

Furthermore, the rarest butterflies occur within ecological systems that harbor other threatened plants and animals that require the same disturbances. The Fender's Blue depends on the threatened Kincaid's Lupine. The St. Francis' Satyr co-occurs with the endangered Rough-leaved Loosestrife and the Red-cockaded Woodpecker, as well as rare plants such as the Venus Flytrap. Like the rarest butterflies, many of these other plants and animals are threatened by vegetative succession to shrubs and trees. Unlike the rarest butterflies, individuals of these species can often survive disturbance. Plants, for example, survive disturbance belowground or survive in a seed bank that remains in the soil. Natural disturbance in these landscapes surely plays a role in restoration of diverse communities of threatened insects, and as the status of these insects is largely unknown, our best strategy to protect them is through restoration for the rarest butterflies.

OF WHAT VALUE ARE THE RAREST BUTTERFLIES?

When I give talks to nonscientific audiences, I am nearly always asked: Why should we care about these rare butterflies? People asking this question assume that, at the very least, the butterflies are pollinators. They are disappointed when I tell them that they are not effective or important for pollination. The rarest butterflies are so few in number and so light in weight that their effects on any ecosystem are negligible. Thus, I face this quandary: the rarest produce no measurable value to human economies or to ecological systems.

Still, the rarest butterflies are valuable. They reveal the diversity of threats to nature, including to other insects that do provide tangible value to people. Although the rarest butterflies do not provide direct value to ecosystems, they are effective indicators of forthcoming loss of other insects. We increasingly hear news about the loss of insect pollinators, insect predators, and other insects that are crucial components of ecological systems. The rarest butterflies do not serve these functions. Because they and other butterflies are so well known, however, they are important indicators of decline for species that do. Whereas the rarest butterflies might disappear without notice by most people, plummeting numbers of other insects will transform ecological systems upon whose functioning we depend.

Another important value of butterflies—more than most other insects—is their visual appeal. They are ambassadors for nature, especially as they educate people about insects and biodiversity. People, even if they cannot name a butterfly species, often tell me about butterflies they observe in their backyards. "I've seen a lot of butterflies this year," they say, or—much more often—"I've noticed many fewer butterflies at my flowers." When they've not seen many butterflies they ask: "Are you and other scientists noticing butterfly decline?" Earlier in my career, I nodded politely in response to backyard reports. Now I think these observations are real and are trending. Butterflies serve as a beacon of biodiversity decline.

Of course, the one species with the highest aesthetic value must be the Monarch. Millions of people enjoy seeing this large, beautiful, and recognizable species. People's willingness to invest time and money to plant milkweed to feed Monarch caterpillars and to plant flowers to attract and feed adults reflects the species' value. The Monarch also holds educational value; as the best known of all butterflies it teaches us about the life cycles and habitats of insects and about the nature of the process of migration. Schoolchildren raise Monarchs from caterpillars to adults, learning the wonders of nature at a very young age. As an herbivore and a pollinator, the Monarch is different from the rarest butterflies in that it provides functional value to ecosystems. However, if Monarch populations continue to fail, the aesthetic costs will be high.

The aesthetic value of Monarchs spills over to other butterflies. For example, I notice in my own research that Monarchs generate support for research and conservation of other pollinators. Efforts targeted at pollinators include more attention to butterflies. In this way, the rarest butterflies that I study ride the long coattails of Monarchs. Because of Monarchs, people have extended their interest to other butterfly species. Interest in butterfly watching, akin to bird watching, is rising in popularity. A growing number of butterfly houses at science centers connect people to a greater diversity of butterflies than they can experience in their neighborhoods or in nearby landscapes. A person visiting natural areas accrues aesthetic value when his/her visits are motivated by a search for unfamiliar butterflies.

Of course, the positive value of the rarest butterflies must be weighed against the costs of saving them. Costs include the land needed for their protection, scientific research needed to determine threats, and restoration actions needed for recovery. The rarest butterflies have mostly persisted on lands that are marginal for agriculture or urbanization. Because of this, there are low-opportunity costs in terms of alternative uses of the land where these butterflies are found. Investment in two types of land acquisition is now needed. First, higher butterfly numbers depend on higher amounts of suitable habitat. Second, at times conservation of land adjacent to habitats is necessary to keep those habitats suitable for the butterfly. One example is the habitat of the endangered Mitchell's Satyr, which lives in fens, peaty wetlands fed by surface water or groundwater. The area of the fen needs protection, but so do areas around the fen where the hydrology degrades after water withdrawals or land-use modification alters water input. For now, new conservation of hundreds of acres will stabilize populations of the rarest butterflies. These levels are well within the range of acreage commonly purchased in modern conservation planning.

Scientific research to understand natural history, population trends, population demography, and population growth in response to habitat degradation and restoration carries a cost. As I've shown, it can take quite a while to discover the intricacies of butterfly biology that are essential to successful conservation.

Significant and enduring costs are borne by those who maintain natural areas that are high in quality for butterflies. As I've discussed, one consistent aspect of restoration for the rarest butterflies is ongoing disturbance. In some cases, recovery of natural disturbances could be self-sustaining. However, because people have worked to eliminate disturbances from the landscape, repeated restoration by people is required (as shown in Figure 10.1). This is especially true where rare butterflies need fire, a disturbance that must happen only under strict control. Fire imposes costs to people who live on private lands near the disturbance. Although escape of intentional and controlled fire is possible, this outcome is extremely rare. For now, rare butterflies including the Fender's Blue and the St. Francis' Satyr are conservation-reliant. This will impose an ongoing cost until natural processes of disturbance reemerge.

MY POSITIVE VISION

What will it take to recover the rarest butterflies? I am an optimist—some might even say an over-optimist—who sees a world of possibilities for rare butterflies. This motivated me to step into the world of the rarest butterflies in the first place, as I was confident that I'd bring science and conservation to stabilize and increase population sizes. While my journey has at times chipped away at my optimism, I have been inspired by the stories of each of the rarest butterflies. These stories have shown that real conservation and recovery can be achieved.

Given that we have low and declining populations of the rarest butterflies, how do we reverse course? As a result of my search, I see four elements of change: investment in the science of natural history, adoption of restoration of whole ecological systems, incorporation of basic ecological principles in land conservation, and harmonizing landscapes occupied by rare butterflies and people. All of these elements presuppose interest by the broad public in conservation of butterflies and pollinators.

A surprising aspect of successful conservation of the rarest butterflies has been discovery in the often-ignored area of basic natural history. This is not sexy science, and we often assume such knowledge is in place, even when it is not. Where research and conservation attention has focused intently on natural history, there have been strong moves to stabilize populations that have been in decline. For example, key elements of conservation worked well for the Schaus' Swallowtail, such as greenhouse rearing. I can make a more ambivalent case for the Large Blue in England; the dedication and persistence of scientists to understand its natural history awes me. The lesson I learn from the Large Blue is that it is never too early for proactive science to flip the trajectory of rare and declining butterflies through an understanding of their natural history.

Butterflies that are most likely to benefit from investment in understanding their natural history are those whose populations have declined precipitously, even though their habitat appears intact. Here I am thinking about widespread species such as the Miami Blue in the 1980s and the Poweshiek Skipperling in the 1990s. It is still not known how these widespread and abundant species were reduced to small population ranges and numbers almost overnight. Large swaths of their habitat appeared to be intact, but some unknown environmental change transformed their environments. Solutions to events like these will put the butterflies, and their ecological systems, on track toward recovery.

Restoration of the rarest butterflies will not happen if the focus is on the rare butterflies alone. They are part of complex, degraded ecological systems. The most positive outcomes for restoration and recovery of the rarest butterflies have come only after the restoration of whole systems. For the Fender's Blue, such an outcome occurred after restoration of grassland systems with fire. For the Bay Checkerspot, it has come after restoration of serpentine grasslands by removal of invasive species. For the St. Francis' Satyr, it has come after restoration of entire wetland ecosystems. In each case, restoration has focused on the ecosystem first. This has had the benefit of restoring degraded systems not only for one butterfly species; many other plants and animals, including rare species, benefit as well.

A theme that is repeated time and again for the rarest butterflies is that landscape conservation must apply basic ecological principles related to the area and interconnection of habitats. As the habitat area available to each species is small, more area is needed. Incremental increases in habitat area will have a noticeably large effect on populations of the rarest butterflies. Another critical element of rare butterfly conservation is a focus on metapopulations. Butterfly population sizes naturally fluctuate between high and low levels. Disturbance magnifies these fluctuations, as it creates better habitats in the future, but at the expense of some individuals. The butterfly population will persist only if butterflies can recolonize disturbed areas, with migration coming from neighboring populations. This will happen only if populations are near enough to permit dispersal. A critical aspect of landscape conservation is to protect nearby areas and landscape corridors that connect populations.

Ultimately, I believe that the rarest butterflies do have a fighting chance for persistence because they can live in harmony with people. Indeed, they will have to. In some cases, butterflies perform best where people act, or should act, in their own best interests. The St. Francis' Satyr might be extinct now if not for intense army activities within artillery ranges. The Crystal Skipper might be extinct if people always built their vacation homes on the edge of the beach (which would make houses vulnerable to hurricane damage). Fender's Blue populations are growing in places where homes and farms are in view.

The rarest butterflies will stand or fall in areas where people change use of the land at the scale of their own land parcels. For many of the rarest butterflies, their populations occupy small areas measured in acres—the size of a house lot or a neighborhood or a farm field. That these areas are so small makes land conservation more feasible, and it brings people close to threatened species. Of course, the proximity to people exposes the rarest butterflies to multiple threats that act in concert to erode their populations. However, this provides opportunities for people to understand and mitigate threat. Again, a key element of conservation in such parcels is a concerted effort to conserve land that reconnects populations.

Butterflies will recover in landscapes occupied by people after multiple parties come to the table. The remarkable reversal of the population decline of the Fender's Blue came about only because of combined and dedicated investment by scientists, conservationists, and land managers. Scientists fine-tuned their studies to address management issues. Conservationists fine-tuned their land management with the best available science. With this combined effort of deliberate and sustained science and conservation, recovery is possible.

In this book, I argue that the list of the rarest butterflies is lengthening and that the threats are becoming more numerous and more intense. However, even against the rising threats, there are signs that the ongoing, downward population trends of the rarest butterflies can be reversed. Saving rare butterflies from extinction can happen only with proper science that enables us to understand butterfly biology and with proper conservation that protects natural places and restores natural function to ecological systems.

CONCLUSION

Saving the rarest butterflies in the world requires one guiding principle: humans should not be the cause of their extinction. This principle extends more broadly, beyond the rarest butterflies to any animal or plant species. With each loss, there is further erosion of the biodiversity of our planet.

My search for the rarest butterflies reflects my interest to conserve them and to develop a more general understanding of many different butterflies. If we can understand threats to the rarest butterflies, we can use that knowledge to conserve more butterflies, more insects, and more biodiversity.

ACKNOWLEDGMENTS

My close friends and colleagues Rob Dunn, Paul Ehrlich, and Alon Tal helped me to see how I could address broader themes of science and conservation through stories of rare butterflies brought together in a book. I am thankful for their support from start to finish.

Students from my lab are the world's experts on three of the butterflies in this book: the St. Francis' Satyr, the Crystal Skipper, and the Miami Blue. Stories of these rare butterflies draw on my students' discoveries related to the rare butterflies they study, other rare butterflies, and rare species in general. Our lab's efforts for twenty years have been led at various times by Brian Hudgens, Daniel Kuefler, Jessica Abbott, Becky Harrison, Allison Leidner, Noa Davidai, Nicole Thurgate, Laura Vogel Milko, Heather Cayton, Johnny Wilson, Elsita Kiekebush, Erica Henry, Erik Aschehoug, Frances Sivakoff, Tyson Wepprich, Jenny McCarty, Ben Pluer, and Victoria Amaral. We accomplished our work with the contributions of about one hundred undergraduate students.

I have benefited from the wisdom of three luminaries in research on and conservation of the rarest butterflies. Since I first met Stu Weiss while I was an undergraduate, he has generously shared his experiences with the Bay Checkerspot. Over the years, I have learned from his successes and his setbacks. I have collaborated with Cheryl Schultz on studies of rare butterfly population biology for a decade, and this has brought me closer to her work on the Fender's Blue. When my mind drifts to more general topics, Cheryl grounds my focus in the critical details of butterfly biology. Jaret Daniels holds two roles that have guided my search for rare butterflies. As the expert in the biology of South Florida's rare butterflies, especially the Miami Blue and the Schaus' Swallowtail, he has shared his wealth of knowledge. As director of one of the world's premiere butterfly centers and collections, he has been a go-to source about butterfly diversity and the world's rarest species.

Many colleagues have broadened my view of the potential for rare butterflies to advance basic ideas in ecology. Elizabeth Crone, Bill Morris, Gina Himes Boor, Brian Hudgens, Allison Louthan, Cheryl Schultz, Jeff Walters, and Norah Warchola have spearheaded collaborations on bigger themes that are applicable to rare butterflies and to plants and animals in general.

Without the efforts of the biologists and land managers responsible for the conservation and recovery of rare butterflies on state and federal land, some of the butterflies in this book might have already gone extinct. These stewards have convinced me that there is no one-way street from science to implementation. Rather, the interaction between science and management goes in both directions, is ongoing, and is necessary. My lab and I have formed close partnerships with Chad Anderson, Brian Ball, Jackie Britcher, Steve Hall, Becky Harrison, Erich Hoffman, Phillip Hughes, Anne Morkill, Jimi Saddle, Mark Salvato, Sarah Steele Cabrera, Tom Wilmers, and Kate Watts. They and others at the US Fish and Wildlife Service, the Florida Keys National Wildlife Refuges Complex, the National Park Service, the Department of Defense, and the Strategic Environmental Research and Development Program have dedicated time and funding toward my lab's efforts and to species conservation.

I received valuable feedback from many people, and they helped me to improve my storytelling. A number of friends and colleagues reviewed the entire book, including Heather Cayton, Julie Doll, Kathryn Haddad, David Wilcove, Cheryl Schultz, Jaret Daniels, and David Pavlik; or sections of the book, including Neal Franks, Betty Franks, Paul Ehrlich, Rob Dunn, Alon Tal, Steve Pousty, Holly Menninger, Stu Weiss, Leslie Ries, Sarah Saunders, Elise Zipkin, Bob Pyle, Allison Leidner, Erica Henry, Matthew Booker, Art Shapiro, Sean Ryan, Jack Liu, Peter Singleton, Karen Oberhauser, Will Wetzel, Paul Severns, and Sarah Steele Cabrera.

Hillary Young, Rodolfo Dirzo, and Anurag Agrawal were generous in allowing me to reproduce figures from their articles. These figures are central to this story and to understanding the science and decline of rare butterflies and rare insects.

While writing, I had the opportunity to admire the work of many talented photographers. They shared their work generously. Thank you to Stu Weiss, Cheryl Schultz, Brian Hudgens, Randy Newman, Molly McCarter, Andy Warren, Jim Brock, Jenny McCarty, Jeff Pippen, Alana Edwards, Jaret Daniels, Kim Davis, Mike Stangeland, Johnny Wilson, Dave Pavlik, Martha Reiskind, Nick Grishin, Elizabeth J. Evans, Peter Law, Bobby McKay, and Donald Gudehus. I wish I had space to include more of their beautiful photos.

I am grateful for Neil McCoy's contributions to figure design and his feedback on photos and layout. He has an uncanny ability to distill complex ideas into graphics that are meaningful and interesting.

My editor at Princeton University Press, Alison Kalett, helped to shape the arc of my story. She endured early drafts, and I was buoyed by her positive spirit. Ellen Foos skillfully ushered the book through production. The manuscript benefited from Amy K. Hughes's thoughtful edits.

My parents, Pat and Nick Haddad, exposed me to nature from an early age, especially during summers at the family farm on the shores of the Chesapeake Bay. They were steadfast in their enthusiasm and interest as I wound my way through studies in ecology and, ultimately, to rare butterflies.

Kathryn, Helen, and Owen have heard many years' worth of stories about the rarest butterflies, often repeatedly. Mostly for better, sometimes for worse, they have accommodated spring break and other trips to research sites where rare butterflies fly. They have each had opportunities to see a few of these rare species. My work and this book are possible only because of them.

NOTES

PREFACE

Page x: George Austin had a wry sense of humor, and in retrospect the name *Inglorius mediocris* is unsurprising; I can imagine George's grin when he named it. *Inglorius mediocris* was not the only new discovery George reported in this paper; there was another new species, *Niconiades incomptus*. Also among the butterflies George described was one species that had previously been found only as far north as southern Brazil, one species known only as far north as Guyana, and two species known only as far north as Peru. This further emphasizes the lack of attention to the rarity of tropical butterflies. Austin, G. T., 1997, "Notes on Hesperiidae in northern Guatemala, with descriptions of new taxa," *Journal of the Lepidopterists' Society* 51:316–332.

Page xi: Butterflies were the sole focus of my dissertation. Haddad, N. M., 1999, "Corridor and distance effects on interpatch movements: A landscape experiment with butterflies," *Ecological Applications* 9:612–622. After I created the experiment with another graduate student, Robert Cheney, we worked with the US Forest Service to attract experts on birds, mammals, plants, pollinators, and other species.

CHAPTER 1. A SLIVER OF CREATION

Page 1: When I began work at Fort Bragg, all known populations of the St. Francis' Satyr had been discovered by Fort Bragg biologist Erich Hoffman and Natural Heritage Program biologist Steve Hall. They discovered twenty-four sites occupied by the St. Francis' Satyr, located on seven different creeks. (I define all individuals on a creek as a population.) Hall, S. P., and E. L. Hoffman, 1994, "Supplement to the rangewide status survey of Saint Francis' Satyr *Neonympha mitchellii francisci* (Lepidoptera: Nymphalidae); 1993 field season," report to the US Fish and Wildlife Service, Region 6 Endangered Species Office, Asheville, NC.

Page 2: As of this writing, Giant Panda population size is 1,864; see Swaisgood, R. R., D. Wang, and F. Wei, 2018, "Panda downlisted but not out of the woods," *Conservation Letters* 11(1):e12355. Black Rhino population size is 5,250; see Emslie, R. H., T. Milliken, B. Talukdar et al., 2011, "African and Asian rhinoceroses—status, conservation and trade: A report from the IUCN Species Survival Commission (IUCN SSC), African and Asian Rhino Specialist Groups, and TRAFFIC to the CITES Secretariat pursuant to Resolution Conf. 9.14 (Rev. CoP15)," United Nations Framework Convention on Climate Change, Conference of the Parties 17, CoP17 doc. 68, annex 5. Northern Spotted Owl population size is 3,000–6,000; see US Fish and Wildlife Service, 2011, "Revised recovery plan for the Northern Spotted Owl (*Strix occidentalis caurina*)," US Department of Interior, Portland, OR.

Page 2: For a synthesis of species numbers across many estimates, see van Nieukerken, E. J., L. Kaila, I. J. Kitching et al., 2011, "Order Lepidoptera Linnaeus, 1758, " in Zhang, Z. Q., ed., "An outline of higher-level classification and survey of taxonomic richness," special issue, *Zootaxa* 3148:212–221. This does not include the number of subspecies, the level of classification that I feature in this book. To put the number of subspecies relative to species in perspective, I have assembled several regional examples: 1) North America north of Mexico, 843 species and 1,008 subspecies, according to Pelham, J. P., 2018, "A catalogue of the butterflies of the United States and Canada," Butterflies of America, rev. Sept. 18, 2018, http://www.butterfliesofamerica.com/US-Can-Cat.htm; 2) United Kingdom, 126 species and 30 subspecies, Agassiz, D.J.L., S. D. Beavan, and R. J. Heckford, 2013, *Checklist of the Lepidoptera of the British Isles*, Handbooks for the Identification of British Insects (St. Albans, UK: Royal Entomological Society); 3) Greece, 235 species and 7 subspecies, "The Butterflies of Greece," users.auth.gr/%7Eefthymia/Butterflies/, accessed Oct. 27, 2018; and 4) Chile, 124 species and 24 subspecies (excluding Lycaenidae), Pyrcz, T. W., A. Ugarte, P. Boyer et al., 2016, "An updated list of the butterflies of Chile (Lepidoptera, Papilionoidea and Hesperioidea) including distribution, flight period and conservation status, pt. 2: Subfamily Satyrinae (Nymphalidae), with the descriptions of new taxa," *Boletín del Museo Nacional de Historia Natural* 65:31–67.

Page 2: For a review of a variety of methods used to estimate insect diversity, the numbers of species they estimate, and this consensus estimate, see Stork, N. E., 2018, "How many species of insects and other terrestrial arthropods are there on earth?," *Annual Review of Entomology* 63:31–45.

Page 2: Henry Bates foreshadowed the important role of butterflies when he wrote, "As the laws of Nature must be the same for all beings, the conclusions furnished by this group of insects must be applicable to the whole organic world; therefore, the study of butterflies—creatures selected as the types of airiness and frivolity—instead of being despised, will some day be valued as one of the most important branches of Biological science." Bates, H. W., 1863, *The Naturalist on the River Amazons*, vol. 2 (London: Murray), 346.

Page 2: This store of knowledge was made possible by the detailed natural history accumulated over two centuries (summarized for North America in Leach, W., 2013, *Butterfly People: An American Encounter with the Beauty of the World* [New York: Pantheon]) and the implementation of citizen science, which engages large groups of nonscientists to collect data on all butterfly species over large regions (see the most prominent example in Pollard, E., and T. J. Yates, 1993, *Monitoring Butterflies for Ecology and Conservation: The British Butterfly Monitoring Scheme* [London: Chapman and Hall]; and the growing number of regional efforts in Taron, D., and L. Ries, 2015, "Butterfly monitoring for conservation," in J. C. Daniels, *Butterfly Conservation in North America* [Dordrecht, Netherlands: Springer], 35–57).

Page 2: In terms of butterflies, the "group" is superfamily Papilionoidea in the order Lepidoptera (butterflies and moths).

Page 2: E. A. MacGregor estimated three billion Painted Lady butterflies in the foothills of the Sierra Nevada, California. To do this, he extrapolated from the forty-mile-wide surveys and the rate at which he observed Painted Ladies flying. McGregor, E. A., 1924, "Painted Lady butterfly (*Vanessa cardui*)," *Insect Pest Survey Bulletin* 4:70. Others have guessed such numbers in California in the past two decades, likely based on this early estimate.

Page 2: I asked Sean Ryan, director of the Pieris Project (www.pierisproject.org/), his rough estimate of the worldwide population size of the Small Cabbage White. He considers it to be the most abundant species on earth, into the billions of individuals. He once attempted to estimate the size using DNA samples and found the population sizes were so large he could not even obtain the low-range estimate.

Page 3: The range of the Miami Blue has contracted from much of the state of Florida to about fifty acres on several tiny islands; see Saarinen, E. V., and J. C. Daniels, 2012, "Using museum specimens to assess historical distribution and genetic diversity in an endangered butterfly," *Animal Biology* 62:337–350.

Page 5: This list also reflects the greater number of subspecies (relative to number of species) in the United States; see Pelham 2018, "A catalogue of the butterflies of the United States and Canada."

Page 5–6: The primary difference in regulation between the threatened and endangered classifications is that regulation for threatened species is more flexible in general, allowing for reduced or expanded protections, and more flexible for natural-resource management within states, especially in regard to "take" (the term used in the ESA for the killing, wounding, trapping, or moving of species).

Page 6: IUCN, 2018, The IUCN Red List of Threatened Species, version 2018-1, http://www.iucnredlist.org.

Page 6: Convention on International Trade in Endangered Species of Wild Fauna and Flora, www.cites.org/eng.

Page 6: United Kingdom Butterfly Monitoring Scheme, www.ukbms.org.

Page 6: North American Butterfly Association, www.naba.org.

Page 7: For one general framework to categorize rarity, with application to plants, see Rabinowitz, D., 1981, "Seven forms of rarity," in H. Synge, ed., *The Biological Aspects of Rare Plant Conservation* (Somerset, NJ: John Wiley and Sons), 205–217.

Page 7: For a concise summary, see Soulé, M. E., and L. S. Mills, 1998, "No need to isolate genetics," *Science* 282:1658–1659.

Page 9: For more detail about how subspecies are determined and the relationship between subspecies and species, see Haig, S. M., E. A. Beever, S. M. Chambers et al., 2006, "Taxonomic considerations in listing subspecies under the US Endangered Species Act," *Conservation Biology* 20:1584–1594.

Page 11: For detailed methods and results of Chris Hamm's research on Mitchell's Satyr genetics, see Hamm, C. A., V. Rademacher, D. A. Landis, and B. L. Williams, 2013, "Conservation genetics and the implication for recovery of the endangered Mitchell's Satyr butterfly, *Neonympha mitchellii mitchellii*," *Journal of Heredity* 105:19–27.

Page 11: Ongoing discussion about the usefulness and consistency of application of the subspecies concept to butterflies is found in Braby, M. F., R. Eastwood, and N. Murray, 2012, "The subspecies concept in butterflies: Has its application in taxonomy and conservation biology outlived its usefulness?," *Biological Journal of the Linnean Society* 106:699–716.

Page 12: A compilation of vertebrate extinctions is found in Ceballos, G., P. R. Ehrlich, A. D. Barnosky et al., 2015, "Accelerated modern human-induced species losses: Entering the sixth mass extinction," *ScienceAdvances* 1(5):e1400253.

Page 13: The Baltimore Checkerspot story was described by Samuel Scudder in Scudder, S. H., 1889, *The Butterflies of the Eastern United States and Canada: With Special Reference to New England*, vol. 1 (Cambridge, MA: printed by author).

Page 15: Herman Behr's letter is housed at the Chicago Field Museum; it is described in Pyle, R. M., 2000, "Resurrection ecology: Bring back the Xerces Blue!," *Wild Earth* 10(3):30–34.

Page 15: The best summary of the butterfy's decline to extinction (before recovery) is presented in Thomas, J., 1980, "Why did the Large Blue become extinct in Britain?" *Oryx* 15:243–247.

Page 17–18: In addition to the information in Figure 1.2, this important review of many studies shows that, worldwide, one-third of rare butterflies are declining in abundance, and that human-caused disturbance reduces butterfly diversity; Dirzo, R., H. S. Young, M. Galetti et al., 2014, "Defaunation in the Anthropocene," *Science* 345:401–406.

Page 18: An important overview of the worldwide threats to butterflies, including effects of habitat loss and fragmentation, is found in Thomas, J. A., 2016, "Butterfly communities under threat," *Science* 353:216–218.

Page 18: This finding is supported by others. One report in Germany showed the loss of 75 percent of insect biomass in a twenty-seven-year time span; Hallmann, C. A., M. Sorg, E. Jongejans et al., 2017, "More than 75 percent decline over 27 years in total flying insect biomass in protected areas," *PloS ONE* 12:e0185809. Another in Loquillo National Forest, Puerto Rico, showed that insect abundances had declined to 25 percent or less in number between 1976 and 2013; Lister, B. C., and A. Garcia, 2018, "Climate-driven declines in arthropod abundance restructure a rainforest food web," *Proceedings of the National Academy of Sciences* 115(44):E10397–E10406, doi.org/10.1073/pnas.1722477115.

CHAPTER 2. BAY CHECKERSPOT

Page 22: For a full synthesis, see Ehrlich, P. R. and I. Hanski, eds., 2004, *On the Wings of Checkerspots: A Model System for Population Biology* (Oxford: Oxford University Press).

Page 22: Ehrlich's first paper on the Bay Checkerspot recognized the importance of spatial structure for its population dynamics; Ehrlich, P. R., 1961, "Intrinsic barriers to dispersal in checkerspot butterfly," *Science* 134:108–109. This paper preceded theoretical development of the metapopulation concept.

Page 22: As summarized in Hanski, I., P. R. Ehrlich, M. Nieminen et al., "Checkerspots and conservation biology," in Ehrlich and Hanski, eds., 2004, *On the Wings of Checkerspots*, 264–287.

Page 23–24: For a detailed history of how people transformed the environment of the San Francisco Bay area over the past two and a half centuries, see Booker, M. M., 2013, *Down by the Bay: San Francisco's History between the Tides* (Berkeley: University of California); pages 23–25 are of particular relevance.

Page 24: Sternitzky described "a race [of butterflies] in the San Francisco Bay district that certainly is entirely different, and probably remained undescribed, due to the fact that so little was known about the species in general." He called it a "variant," and it was later described as a subspecies. Sternitzky, R., 1937, "A race of *Euphydryas editha* BDV. (Lepidoptera)," *Canadian Entomologist* 69:203–205.

Page 26: For a detailed history of population loss since the 1950s, see Murphy, D. D., and S. B. Weiss, 1988, "Ecological studies and the conservation of the Bay

Checkerspot butterfly, *Euphydryas editha bayensis*," *Biological Conservation* 46:183–200.

Page 27–28: See Weiss, S. B., 1999, "Cars, cows, and checkerspot butterflies: Nitrogen deposition and management of nutrient-poor grasslands for a threatened species," *Conservation Biology* 13:1476–1486.

Page 29: See Kuussaari, M., J. Van Nouhys, J. J. Hellmann, M. C. Singer, "Larval biology of checkerspots," in Ehrlich and Hanski, eds., 2004, *On the Wings of Checkerspots*, 138–160.

Page 29: See Ehrlich, P. R., D. D. Murphy, M. C. Singer et al., 1980, "Extinction, reduction, stability and increase: The responses of checkerspot butterfly (*Euphydryas*) populations to the California drought," *Oecologia* 46:101–105.

Page 30–31: See Weiss, S. B., D. D. Murphy, and R. R. White, 1988, "Sun, slope, and butterflies: Topographic determinants of habitat quality for *Euphydryas editha*," *Ecology* 69:1486–1496.

Page 32: For a concise overview of the metapopulation concept, especially as it applies to conservation, see Hanski, I., and D. Simberloff, "The metapopulation approach, its history, conceptual domain, and application to conservation," in Hanksi, I., and M. E. Gilpin, 1997, *Metapopulation Biology: Ecology, Genetics, and Evolution* (San Diego, CA: Academic Press), 5–26. The concept was applied to the Bay Checkerspot by Susan Harrison in Harrison, S., D. D. Murphy, and P. R. Ehrlich, 1988, "Distribution of the Bay Checkerspot butterfly, *Euphydryas editha bayensis*: Evidence for a metapopulation model," *American Naturalist* 132:360–382.

Page 34–36: For a broad overview of the restoration efforts and Bay Checkerspot population responses, see Neiderer, C., 2018, "Bay Checkerspot Reintroduction: Coyote Ridge to Edgewood Natural Preserve," report of the Creekside Center for Earth Observation, Menlo Park, CA.

Page 38: See Ehrlich, P. R., and D. D. Murphy, 1987, "Conservation lessons from long-term studies of checkerspot butterflies," *Conservation Biology* 1:122–131.

CHAPTER 3. FENDER'S BLUE

Page 41: This is the first of three chapters on the group of butterflies referred to as the "blues." In 1993, Hall Cushman and Dennis Murphy wrote: "We suspect that [blue butterflies] possess biological characteristics that put them at special risk, and may predispose them to extinction"; see Cushman, J. H., and D. D. Murphy, 1993, "Susceptibility of Lycaenid butterflies to endangerment," *Wings* 17:16–21.

Page 41: See Macy, R. W., 1931, "A new Oregon butterfly (Lepid. Lycaenidae)," *Entomological News* 42:1–2.

Page 42: The extent of Pacific Northwest prairie and other features of Fender's Blue habitat are summarized in US Fish and Wildlife Service, 2006, "Endangered and threatened wildlife and plants: Designation of critical habitat for the Fender's Blue butterfly (*Icaricia icarioides fenderi*), *Lupinus sulphureus* ssp. *kincaidii* (Kincaid's Lupine), and *Erigeron decumbens* var. *decumbens* (Willamette Daisy)"; final rule, *Federal Register* 71:63862–63977.

Page 42–43: For broad historical analyses of landscape change in the Willamette River valley, see Towle, J. C., 1982, "Changing geography of Willamette Valley woodlands," *Oregon Historical Quarterly* 83:66–87; and Johannessen, C. L., W. A. Davenport, A. Millet, and S. McWilliams, 1971, "The vegetation of the Willamette Valley woodlands," *Annals of the Association of American Geographers* 61:286–302.

Page 44–45: For a very accessible account of the Fender's Blue's history, see Schultz C. B., 2015, "Flying towards recovery: Conservation of Fender's Blue butterfly," *News of the Lepidopterists' Society* 57:210–213.

Page 46: See Smith, C. P., 1924, "Studies in the genus *Lupinus*-XI. Some new names and combinations," *Bulletin of the Torrey Botanical Club* 51:303–310.

Page 50: See US Fish and Wildlife Service, 2010, *Recovery Plan for the Prairie Species of Western Oregon and Southwestern Washington* (Portland, OR: US Fish and Wildlife Service).

Page 50: For details of scientific studies that were used to improve nursery programs, see Severns, P. M., 2003, "Propagation of a long-lived and threatened prairie plant, *Lupinus sulphureus* ssp. *kincaidii*," *Restoration Ecology* 11:334–342.

Page 51: See Schultz, C. B., 2001, "Restoring resources for an endangered butterfly," *Journal of Applied Ecology* 38:1007–1019.

Page 51–52: Restoration actions vary in their effectiveness at increasing butterfly populations and their ease of implementation, and these tradeoffs are summarized nicely in Schultz, C. B., E. Henry, A. Carleton et al., 2011, "Conservation of prairie-oak butterflies in Oregon, Washington, and British Columbia," *Northwest Science* 85:361–388; and Stanley, A. G., P. W. Dunwiddie, and T. N. Kaye, 2011, "Restoring invaded Pacific Northwest prairies: Management recommendations from a region-wide experiment," *Northwest Science* 85:233–246.

Page 53: Many ideas in this chapter on science and conservation trace back to Cheryl Schultz's seminal dissertation work, which included studies of natural history, demography, dispersal behavior, and restoration via burning; Schultz, C. B., 1998, "Ecology and Conservation of Fender's Blue Butterfly" (PhD diss., University of Washington, Seattle).

Page 53: Data provided to the author by C. B. Schultz on Aug. 18, 2018.

Page 54: With information on the sizes of populations over time, including average size and fluctuations in size, scientists can project the size in the future. From this, they can determine the risk of extinction. Conservation biologists use this assessment of risk to prioritize investment in conservation and restoration. See Schultz, C. B., and P. C. Hammond, 2003, "Using population viability analysis to develop recovery criteria for endangered insects: Case study of the Fender's Blue butterfly," *Conservation Biology* 17:1372–1385.

Page 54: For discussion of the balance between the detailed science and practical conservation of the Fender's Blue, see Schultz, C. B., and E. E. Crone, 2015, "Using ecological theory to develop recovery criteria for an endangered butterfly," *Journal of Applied Ecology* 52:1111–1115.

Page 55–57: See Schultz, C. B., A. M. Franco, and E. E. Crone, 2012, "Response of butterflies to structural and resource boundaries," *Journal of Animal Ecology* 81:724–734.

Page 59: See Warchola, N., C. Bastianelli, C. B. Schultz, and E. E. Crone, 2015, "Fire increases ant-tending and survival of the Fender's blue butterfly larvae," *Journal of Insect Conservation* 19:1063–1073.

Page 59–60: See Schultz, C. B., and E. E. Crone, 1998, "Burning prairie to restore butterfly habitat: A modeling approach to management tradeoffs for the Fender's Blue," *Restoration Ecology* 6:244–252.

Page 60: Data provided to the author by C. B. Schultz on Aug. 18, 2018.

Page 62: See Schultz, C. B., and E. E. Crone, 2005, "Patch size and connectivity thresholds for butterfly habitat restoration," *Conservation Biology* 19:887–896.

CHAPTER 4. CRYSTAL SKIPPER

Page 65–66: A history of Crystal Skipper taxonomy beginning with its first capture is chronicled in Burns, J. M., 2015, "Speciation in an insular sand dune habitat: *Atrytonopsis* (Hesperiidae: Hesperiinae)—mainly from the southwestern United States and Mexico—off the North Carolina coast," *Journal of the Lepidopterists' Society* 69:275–292; and Burns, J. M., 2000, "A striking new species of skipper butterfly on the North Carolina coast," 51st Annual Meeting of the Lepidopterists' Society, Wake Forest University, Winston-Salem, NC.

Page 68: See Leidner, A. K., 2009, "Butterfly conservation in fragmented landscapes" (PhD diss., North Carolina State University, Raleigh), http://www.lib.ncsu.edu/resolver/1840.16/5140.

Page 75–76: Allison Leidner created a three-pronged approach that could be used to create a complete understanding of Crystal Skipper populations in fragmented landscapes. She used behavioral studies to determine whether butterflies would venture into non-habitat. If they would not, landscape corridors would be required to connect habitats. She then used mark-and-recapture studies to determine how many butterflies moved between patches. Finally, she used genetic studies to determine whether dispersing individuals established in new areas. If they moved through fragmented landscapes and established populations, then so-called stepping-stone habitats would promote conservation. Leidner, A. K., and N. M. Haddad, 2011, "Combining measures of dispersal to identify conservation strategies in fragmented landscapes," *Conservation Biology* 25:1022–1031.

Page 77: See Leidner, A. K., and N. M. Haddad, 2006, "Behavior of a rare butterfly in natural and urbanized areas: Implications for dune conservation management," report to North Carolina Sea Grant, Raleigh.

Page 78: See Leidner, A. K., and N. M. Haddad, 2010, "Natural, not urban, barriers define population structure for a coastal endemic butterfly," *Conservation Genetics* 11:2311–2320.

Page 81: See Hay, C. C., E. Morrow, R. E. Kopp, and J. X. Mitrovica, 2015, "Probabilistic reanalysis of twentieth-century sea-level rise," *Nature* 517: 481–484.

Page 81: See Kopp, R. E., B. P. Horton, A. C. Kemp, and C. Tebaldi, 2015, "Past and future sea-level rise along the coast of North Carolina, USA," *Climatic Change* 132:693–707.

Page 81: See "Surging Seas Risk Finder, North Carolina, USA," Surging Seas/Climate Central, riskfinder.climatecentral.org/state/north-carolina.us, accessed Sept. 10, 2018.

Page 81–82: Managed relocation, also called "assisted migration," has been the subject of ongoing discussion about its practicality, legality, and ethics. See Schwartz, M. W., J. J. Hellmann, J. M. McLachlan et al., 2012, "Managed relocation: Integrating the scientific, regulatory, and ethical challenges," *Bio Science* 62:732–743; and Hoegh-Guldberg, O., L. Hughes, S. McIntyre et al., 2008, "Assisted colonization and rapid climate change," *Science* 321:345–346.

Page 83: See Burns 2015, "Speciation in an insular sand dune habitat."

CHAPTER 5. MIAMI BLUE

Page 85: See Bethune-Baker, G. T., 1916, "Notes on the genus *Hemiargus* Hübner in Dyer's list (Lep.)," *Entomological News* 27:449–457.

Page 86: For a broad overview of Nabokov's contributions to taxonomy of butterflies, including his detailed drawings used to define species, and connections between his science and literature, see Blackwell, S. H., and K. Johnson, 2016, *Fine Lines: Vladimir Nabokov's Scientific Art* (New Haven, CT: Yale University Press).

Page 86: Klots, A. B., *A Field Guide to the Butterflies of North America, East of the Great Plains* (Boston: Houghton Mifflin Company, 1951).

Page 89: Official estimates of the US National Center for Environmental Information of the National Oceanic and Atmospheric Association; see NOAA Centers for Environmental Information, 2018, "U.S. Billion-Dollar Weather and Climate Disasters, 1980–2018," www.ncdc.noaa.gov/billions/events.pdf, accessed Sept. 10, 2018.

Page 89: Calhoun, J. V., J. R. Slotten, and M. H. Salvato, 2000, "The rise and fall of tropical blues in Florida: *Cyclargus ammon* and *Cyclargus thomasi bethunebakeri* (Lepidoptera: Lycaenidae)," *Holarctic Lepidoptera* 7:13–20.

Page 90: In addition to their use in assembling distributions, museum records provided sources of genetic diversity throughout the Miami Blue's range. See Saarinen, E. V., and J. C. Daniels, 2012, "Using museum specimens to assess historical distribution and genetic diversity in an endangered butterfly," *Animal Biology* 62:337–350.

Page 90: See North American Butterfly Association, "Saving South Florida's butterflies: Miami Blue fund," www.naba.org/miamiblue.html, accessed Sept. 10, 2018.

Page 91: Two areas connected via dispersal, and the resulting gene flow maintained higher than expected genetic diversity; Saarinen, E. V., J. C. Daniels, and J. E. Maruniak, 2014, "Local extinction event despite high levels of gene flow and genetic diversity in the federally-endangered Miami blue butterfly," *Conservation Genetics* 15:811–821.

Page 91: Daniels, J. C., 2010, "Conservation and field surveys of the endangered Miami Blue butterfly (*Cyclargus thomasi bethunebakeri*) (Lepidoptera: Lycaenidae)," report 3, submitted to United States Fish and Wildlife Service, Florida Keys National Wildlife Refuges.

Page 92: For a detailed summary of observations at Bahia Honda and the Marquesas and releases of captive-reared individuals, see US Fish and Wildlife Service, 2012, "Endangered and threatened wildlife and plants; emergency listing of the Miami Blue butterfly as endangered throughout its range; listing of the Cassius Blue, Ceraunus Blue, and Nickerbean Blue butterflies as threatened due to similarity of appearance to the Miami Blue in coastal south and central Florida," *Federal Register* 77(67):20948–20986.

Page 95: See Cannon, P., T. Wilmers, and K. Lyons, 2010, "Discovery of the imperiled Miami Blue butterfly (*Cyclargus thomasi bethunebakeri*) on islands in the Florida Keys National Wildlife Refuges, Monroe County," *Southeastern Naturalist* 9:847–853.

Page 103–5: See Henry, E. H., N. M. Haddad, J. Wilson et al., 2015, "Point-count methods to monitor butterfly populations when traditional methods fail: A case study with Miami Blue butterfly," *Journal of Insect Conservation* 19: 519–529.

CHAPTER 6. ST. FRANCIS' SATYR

Page 116: See Parshall, D. K., and T. W. Kral, 1989, "A new subspecies of *Neonympha mitchellii* (French) (Satyridae) from North Carolina," *Journal of the Lepidopterists' Society* 43:114–119. Given the known population size at the time, it was reasonable to write, "There is not a more endangered butterfly population in the eastern US than *N. m. francisci*." Even with the discovery of additional populations, the authors were not far off the mark.

Page 117–18: For a detailed discussion of illegal collection of the St. Francis' Satyr and other butterflies, see Laufer, P., 2010, *Dangerous World of Butterflies: The Startling Subculture of Criminals, Collectors, and Conservationists* (New York: Lyons Press). See also Alexander, C., "Crimes of passion: A glimpse into the covert world of rare butterfly collecting," *Outside Magazine*, May 2, 2004.

Page 119: US Fish and Wildlife Service, 1995, "Endangered and threatened wildlife and plants; Saint Francis' Satyr determined to be endangered," *Federal Register* 60(17):5264–5267.

Page 120: Other early studies were on larval feeding, dispersal, and flight periods through the year. See Kuefler, D., N. M. Haddad, S. Hall et al., 2008, "Distribution, population structure and habitat use of the endangered St. Francis' Satyr butterfly, *Neonympha mitchellii francisci*," *American Midland Naturalist* 159:298–320.

Page 124: See Taron, D., and L. Ries, "Butterfly monitoring for conservation," in J. C. Daniels, ed., *Butterfly Conservation in North America* (Dordrecht, Netherlands: Springer, 2015), 35–57.

Page 125: In addition to mark-and-recapture and transect count methods, I investigated another technique that generated numbers of individuals per generation based on numbers counted per day through the flight period. All three methods produced results that were highly correlated. Haddad, N. M., B. Hudgens, C. Damiani et al., 2008, "Determining optimal population monitoring for rare butterflies," *Conservation Biology* 22:929–940.

Page 128: See Wilson, J. W., J. O. Sexton, R. T. Jobe, and N. M. Haddad, 2013, "The relative contribution of terrain, land cover, and vegetation structure indices to species distribution models," *Biological Conservation* 164:170–176.

Page 130: For details on the methods used for the St. Francis' Satyr, and for application to other butterfly species, see Cayton, H. L., N. M. Haddad, K. Gross et al., 2015, "Do growing degree days predict phenology across butterfly species?," *Ecology* 96:1473–1479.

Page 130: I first saw the phrase "killing butterflies to save butterflies" used by Larry Orsak in a completely different context. I adopted the phrase in reference to the effects of disturbance; he referred to farming Queen Alexandra's Birdwing (*Ornithoptera alexandrae*) for sale to museums and collectors, which could save butterflies by preventing their harvest in the wild. Orsak, L. J., 1993, "Killing butterflies to save butterflies: A tool for tropical forest conservation in Papua New Guinea," *News of the Lepidopterists' Society* 71–80.

Page 132: See Frost, C. C., J. Walker, and R. K. Peet, "Fire-dependent savannas and prairies of the Southeast: Original extent, preservation status, and management problems," in D. L. Kulhavy and R. N. Connor, eds., *Wilderness and Natural Areas in the Eastern United States: A Management Challenge* (Nacogdoches, TX: Center for Applied Studies, School of Forestry, Stephen F. Austin State University, 1986), 348–357.

Page 133: The St. Francis' Satyr's close relative, the Mitchell's Satyr, benefits from activities of the beaver, but there are other factors that maintain the fen habitat where it is found.

Page 136–41: For an overview of our process from restoration to reintroduction to population growth, see Cayton, H., N. M. Haddad, B. Ball et al., 2015, "Habitat restoration as a recovery tool for a disturbance-dependent butterfly, the endangered St. Francis' Satyr," in Daniels, ed., *Butterfly Conservation in North America* (Dordrecht: Springer), 147–159.

CHAPTER 7. SCHAUS' SWALLOWTAIL

Page 144: Williams, L. K., 1983 (revised by P. S. George, 1995), "South Florida: A Brief History," Historical Museum of Southern Florida, web.archive.org/web/20100429002717/http://www.hmsf.org/history/south-florida-brief-history.htm, accessed Nov. 9, 2018. US Census Office, 1901, Census Reports, vol. 1, 12th Census of the United States (1900).

Page 144: See Schaus, W., 1911, "A new *Papilio* from Florida, and one from Mexico (Lepid.)," *Entomological News* 22:438–439.

Page 147: See Covell, C. V., 1977, "Project Ponceanus and the status of the Schaus swallowtail (*Papilio aristodemus ponceanus*) in the Florida Keys," *Atala* 5:4–6.

Page 148: The 1935 hurricane's record intensity measured as central pressure; see https://www.nhc.noaa.gov/outreach/history/#keys.

Page 148–49: Grimshawe's beautifully written story concludes, "Lower Matecumbe, long a naturalist's paradise with its exotic plants and children of the sun, is like a stricken animal, panting at bay ere it dies. Matecumbe, indeed, is a 'place of sorrow.'" Grimshawe, F. M., 1940, "Place of sorrow," *Nature Magazine* 33:565–567, 611.

Page 149: See Henderson, W., 1945, "*Papilio aristodemus ponceana* Schaus (Lepidoptera: Papilionidae)," *Entomological News* 56:29–32; Henderson, W., 1945, "Additional notes on *Papilio aristodemus ponceana* Schaus (Lepidoptera: Papilionidae)," *Entomological News* 56:187–188; Henderson, W., 1946, "*Papilio aristodemus ponceana* Schaus (Lepidoptera: Papilionidae) Notes," *Entomological News* 57:100–101.

Page 150: See Young, F. N., 1956, "Notes on collecting Lepidoptera in southern Florida," *Lepidopterists' News* 9:204–212.

Page 150: See Klots, A. B., 1951, *A Field Guide to the Butterflies of North America, East of the Great Plains* (Boston: Houghton Mifflin).

Page 150–51: See Rutkowski, F., 1971, "Observations on *Papilio aristodemus ponceanus* (Papilionidae)," *Journal of the Lepidopterists' Society* 25:126–136; Covell, C. V., and G. W. Rawson, 1973, "Project Ponceanus: A report of the first efforts to survey and preserve the Schaus Swallowtail (Papilionidae) in southern Florida," *Journal of the Lepidopterists' Society* 27:206–210; Brown, L. N., 1973, "Populations of *Papilio andraemon bonhotei* Sharpe and *Papilio aristodemus ponceanus* Schaus (Papilionidae) in Biscayne National Monument, Florida," *Journal of the Lepidopterists' Society* 27:136–140.

Page 151: Covell and Rawson also describe the small amount of development they observed on Key Largo, the key closest to Miami; Covell and Rawson 1973, "Project Ponceanus," 206–210.

Page 152: Remarkably, in the last paragraph of his paper Rutkowski describes perfectly the dynamics of a Schaus' Swallowtail metapopulation; Rutkowski 1971, "Observations on *Papilio aristodemus ponceanus*," 136.

Page 153: See Pyle, R. M., 1976, "Conservation of Lepidoptera in the United States," *Biological Conservation* 9:55–75.

Page 153: See Loftus, W. F., and J. A. Kushlan, 1984, "Population fluctuations of the Schaus Swallowtail (Lepidoptera: Papilionidae) on the islands of Biscayne Bay, Florida, with comments on the Bahaman Swallowtail," *Florida Entomologist* 67:277–287.

Page 153–54: See Covell 1977, "Project Ponceanus."

Page 157: See Webster, P. J., G. J. Holland, J. A. Curry, and H.-R. Chang, 2005, "Changes in tropical cyclone number, duration, and intensity in a warming environment," *Science* 309:1844–1846.

Page 158–59: See Eliazar, P. J., and T. C. Emmel, 1991, "Adverse impacts to nontarget insects," in T. C. Emmel and J. C. Tucker, eds., *Mosquito Control Pesticides: Ecological Impacts and Management Alternatives, Conference Proceedings* (Gainesville, FL: Scientific Publishers), 17–19.

Page 159–60: Recent examples include Hammond, A., R. Galizi, K. Kyrou et al., 2016, "A CRISPR-Cas9 gene drive system targeting female reproduction in the malaria mosquito vector *Anopheles gambiae*," *Nature Biotechnology* 34:78–83; and Gantz, V. M., N. Jasinskiene, O. Tatarenkova et al., 2015, "Highly efficient Cas9-mediated gene drive for population modification of the malaria vector mosquito *Anopheles stephensi*," *Proceedings of the National Academy of Sciences* 112:E6736–E6743.

Page 160: Innovation in this area is accelerating, especially since the advent of CRISPR technologies that permit targeted gene editing.

Page 160–61: US Fish and Wildlife Service, 1976, "Determination that two species of butterflies are threatened species and two species of mammals are endangered species," *Federal Register* 41(83):17736–17740.

Page 161: With mounting threats and declining populations, the US Fish and Wildlife Service elevated the Schaus' Swallowtail to endangered in 1984.

Page 160–61: See Pyle, R. M., 1976, "Conservation of Lepidoptera in the United States."

CHAPTER 8. THE FINAL FLIGHT OF THE BRITISH LARGE BLUE

Page 174: See Wynhoff, I., 1998, "The recent distribution of the European *Maculinea* species," *Journal of Insect Conservation* 2:15–27.

Page 174: See Howarth, T. G., 1973, "The conservation of the Large Blue butterfly (*Maculinea arion* L.) in West Devon and Cornwall," *Proceedings and Transactions of the British Entomological and Natural History Society*, 121–126.

Page 174: This chapter draws heavily on the most comprehensive history of the British Large Blue's conservation and decline toward extinction: Thomas, J. A., 1980, "Why did the Large Blue become extinct in Britain?," *Oryx* 15:243–247.

Page 174–75: See Goss, H., 1884, "On the probable extinction of *Lycaena arion* in Britain," *Entomologist's Monthly* 21:107–109; and Marsden, H., 1884, "On the probable extinction of *Lycaena arion* in England," *Entomologist's Monthly* 21:186–189.

Page 175: See Sheldon, W. G., 1925, "The destruction of British butterflies," *Entomologist* 58:105–112.

Page 176: See Frohawk, F. W., 1906, "Completion of the life-history of *Lycaena arion*," *Entomologist* 39:145–147.

Page 176: See Chapman, T. A., 1915, "What the larva of *Lycaena arion* does during its last instar," *Transactions of the Entomological Society of London* 1915:291–312.

Page 176–78: See Purefoy, E. B., 1953, "An unpublished account of experiments carried out at East Farleigh, Kent in 1915 and subsequent years on the life history of *Maculinea arion*, the Large Blue butterfly," *Proceedings of the Royal Entomological Society of London Series A* 28:160–162; and Frohawk, F. W., 1915, "Further observations of the last stage of the larva of *Lycaena arion*," *Transactions of the Entomological Society of London* 1915:313–316.

Page 178: See Thomas, J. A., and J. C. Wardlaw, 1992, "The capacity of a *Myrmica* ant nest to support a predacious species of *Maculinea* butterfly," *Oecologia* 91:101–109.

Page 179: See Patricelli, D., F. Barbero, A. Occhipinti et al., 2015, "Plant defences against ants provide a pathway to social parasitism in butterflies," *Proceedings of the Royal Society of London B: Biological Sciences* 282:20151111.

Page 180: See Thomas 1980, "Why did the Large Blue become extinct in Britain?"

Page 181: See Thomas, J. A., D. J. Simcox, J. C. Wardlaw et al., 1998, "Effects of latitude, altitude, and climate on the habitat and conservation of the endangered butterfly *Maculinea arion* and its *Myrmica* ant hosts," *Journal of Insect Conservation* 2:39–46.

Page 182–83: See Thomas 1980, "Why did the Large Blue become extinct in Britain?"

Page 182–83: See Thomas, J. A., D. J. Simcox, R. T. Clarke, 2009, "Successful conservation of a threatened *Maculinea* butterfly," *Science* 325:80–83.

CHAPTER 9. MONARCHS: THE PERILS FOR ABUNDANT BUTTERFLIES

Page 187: For a broad and accessible overview of Monarch biology, I recommend Agrawal, A., 2017, *Monarchs and Milkweed: A Migrating Butterfly, a Poisonous Plant, and Their Remarkable Story of Coevolution* (Princeton, NJ: Princeton University Press).

Page 188: The area of up to fifty acres where Monarchs roost is protected in the 100,000-acre Monarch Butterfly Biosphere Reserve.

Page 188–91: For methods to estimate population density, see Calvert, W. H., 2004, "Two methods estimating overwintering Monarch population size in Mexico," in K. S. Oberhauser and M. J. Solensky, eds., *The Monarch Butterfly: Biology and Conservation* (Ithaca, NY: Cornell University Press), 121–127. For the generally agreed-upon standard density of 50 million/hectare, see Brower, L. P., D. R. Kust, E. Rendon-Salinas et al., 2004, "Catastrophic winter storm mortality of Monarch butterflies in Mexico during January 2002," in Oberhauser and Solensky, eds., *The Monarch Butterfly*, 151–166.

Page 190: See Walsh, B. D., and C. V. Riley, "A swarm of butterflies," *American Entomologist*, Sept. 1868, 28–29; for descriptions of "great swarms" of Monarchs in New Jersey, see Holland, W. J., 1908, *The Butterfly Book: A Popular Guide to a Knowledge of the Butterflies of North America* (Garden City, NY: Doubleday, Page, and Company), 82–83.

Page 191–92: See Schultz, C. B., L. M. Brown, E. Pelton, and E. E. Crone, 2017, "Citizen science monitoring demonstrates dramatic declines of monarch butterflies in western North America," *Biological Conservation* 214:343–346.

Page 193: Other threats that I do not discuss include climate change and disease.

Page 193: Although these patterns are remarkably well correlated, they don't explain the mechanism of broad-scale decline. See Agrawal 2017, *Monarchs and Milkweed*.

Page 193: John Pleasants and Karen Oberhauser found that between 1999 and 2010 the density of milkweed in agriculture declined 80 percent, three times faster than that in nonagricultural lands. The decline has not abated, as herbicide-tolerant soybeans are on rotation in 94 percent of corn and soybean fields, and the coverage is rising. The area where milkweed is affected extends across a massive range, to over 100 million acres. Pleasants, J. M., and K. S. Oberhauser, 2013, "Milkweed loss in agricultural fields because of herbicide use: Effect on monarch butterfly population," *Insect Conservation and Diversity* 6:135–144.

Page 193: One study tested the effects of neonicotinoids on survival of Monarchs in the lab and of the concentration in milkweeds adjacent to crops; Pecenka, J. R., and J. G. Lundgren, 2015, "Non-target effects of clothianidin on monarch butterflies," *Science of Nature* 102:19. Neonicotinoids in seed coatings are lost during planting as dust that can drift over large areas, exposing insects to their effects well beyond field margins; Krupke, C. H., J. D. Holland, E. Y. Long, and B. D. Eitzer, 2017. "Planting of neonicotinoid-treated maize poses risks for honey bees and other non-target organisms over a wide area without consistent crop yield benefit," *Journal of Applied Ecology* 54:1449–1458.

Page 194: An important and panoramic perspective on change in forest cover, including in the eastern United States, is Williams, M., 2003, *Deforesting the Earth: From Prehistory to Global Crisis* (Chicago: Chicago University Press).

Page 194: The effects of landscape change on Monarch populations depend on their location in their breeding range prior to return to their overwintering sites. University of Maryland biologist Tyler Flockhart identified Monarch breeding regions based on samples taken from individual Monarchs in their overwintering sites. He studied rare elements that Monarchs acquired prior to their journey south. He reasoned that isotopes of hydrogen and carbon varied throughout North America along gradients of latitude, elevation, temperature, and precipitation. Using Monarchs collected in Mexico, he could then match the butterflies to their region of origin. In this way, he found that nearly equal proportions originated in the Midwest as in the Northeast. Flockhart, D. T., L. I. Wassenaar, T. G. Martin et al., 2013, "Tracking multigenerational colonization of the breeding grounds by monarch butterflies in eastern North America," *Proceedings of the Royal Society of London B: Biological Sciences* 280:20131087.

Page 194: See Vidal, O., J. López-García, and E. Rendón-Salinas, 2014, "Trends in deforestation and forest degradation after a decade of monitoring in the Monarch Butterfly Biosphere Reserve in Mexico," *Conservation Biology* 28:177–186.

Page 194: Forest is lost through means other than tree harvest. One threat is potential buildup of mining next to the forest. Older, currently unused mines have tunnels dug through mountains under Monarch overwintering sites. Rather than a direct effect on Monarchs, processing ore sucks up water, and water withdrawal would dry soils and kill Oyamel Fir trees. UNESCO World Heritage Centre, 2017, "State of Conservation: Monarch Butterfly Biosphere Reserve (Mexico)," whc.unesco.org/en/soc/3559, accessed Sept. 13, 2018.

Page 195: Monarch Watch, www.monarchwatch.org.

NOTES TO CHAPTER 9   237

Page 195: Monarch Larva Monitoring Project, University of Minnesota, monarchlab.org/mlmp.

Page 195: North American Butterfly Association, "Butterfly Counts," www.naba.org/butter_counts.html.

Page 195–96: Members of my lab have used the Ohio Lepidopterist Society's data to determine the effects of climate and urban warming on the seasons of butterfly flight. One example is Cayton, H. L., N. M. Haddad, K. Gross et al., 2015, "Do growing degree days predict phenology across butterfly species?," *Ecology* 96:1473–1479.

Page 196: See Zipkin, E. F., L. Ries, R. Reeves et al., 2012, "Tracking climate impacts on the migratory Monarch butterfly," *Global Change Biology* 18:3039–3049.

Page 196: See Ries, L., D. J. Taron, and E. Rendón-Salinas, 2015, "The disconnect between summer and winter monarch trends for the eastern migratory population: Possible links to differing drivers," *Annals of the Entomological Society of America* 108:691–699; and Ries, L., D. J. Taron, E. Rendón-Salinas, K. S. Oberhauser, D. Taron et al., 2015, "Connecting eastern Monarch population dynamics across their migratory cycle," in K. S. Oberhauser, K. R. Nail, S. Altizer, eds., *Monarchs in a Changing World: Biology and Conservation of an Iconic Insect* (Ithaca, NY: Cornell University Press), 268–281.

Page 196: See Saunders, S. P., L. Ries, N. Neupane et al., "Multi-scale factors drive the size of winter Monarch colonies" (unpublished manuscript, in review).

Page 196: Another study, which uses NABA data from throughout the breeding range, has identified the autumn migration range as a possible region that harbors the cause of Monarch decline; Inamine, H., S. P. Ellner, J. P. Springer, and A. A. Agrawal, 2016, "Linking the continental migratory cycle of the Monarch butterfly to understand its population decline," *Oikos* 125:1081–1091.

Page 196: See Caldwell, W., C. L. Preston, and A. Cariveau, 2018, *Monarch Conservation Implementation Plan* (St. Paul, MN: Monarch Joint Venture), monarchjointventure.org/images/uploads/documents/2018_Monarch_Conservation_Implementation_Plan_FINAL_2.pdf.

Page 198: Pollinator Health Task Force, June 2016, *Pollinator Partnership Action Plan* (Washington, DC: The White House), https://www.whitehouse.gov/sites/whitehouse.gov/files/images/Blog/PPAP_2016.pdf.

Page 198: For example, Haddad, N. M., D. R. Browne, A. Cunningham et al., 2003, "Corridor use by diverse taxa," *Ecology* 84:609–615.

Page 199: See Ries, L., D. M. Debinski, and M. L. Wieland, 2001, "Conservation value of roadside prairie restoration to butterfly communities," *Conservation Biology* 15:401–411.

Page 200: Extinction of the Rocky Mountain Locust (*Melanoplus spretus*) followed a trajectory similar to that of the Passenger Pigeon. The locust once inhabited all of North America and especially prairies. In 1875, a farmer extrapolated his observations across the locust's range and estimated ten trillion individuals. As farmers plowed prairies, they simultaneously degraded locust habitat. By 1903 the Rocky Mountain Locust was extinct. Lockwood, J. A., 2009, *Locust: The Devastating Rise and Mysterious Disappearance of the Insect That Shaped the American Frontier* (New York: Basic Books).

Page 200–201: See Homer-Dixon, T., 2014/2018, "Today's butterfly effect is tomorrow's trouble," *Globe and Mail* (Toronto), Nov. 15, 2014, updated May 12, 2018.

CHAPTER 10. THE LAST BUTTERFLY?

Page 203: See Agrawal, A. A., and H. Inamine, 2018, "Mechanisms behind the monarch's decline," *Science* 360:1294–1296.

Page 203: See Dirzo, R., H. S. Young, M. Galetti et al., 2014, "Defaunation in the Anthropocene," *Science* 345:401–406.

Page 203: See Hallmann, C. A., M. Sorg, E. Jongejans et al., 2017, "More than 75 percent decline over 27 years in total flying insect biomass in protected areas," *PLoS ONE* 12(10):e0185809.

Page 205: See Schultz, C. B., N. M. Haddad, E. H. Henry, and E. E. Crone, forthcoming (2019), "Movement and demography of at-risk butterflies: Building blocks for conservation," *Annual Review of Entomology* 64: 167–184.

Page 206: I must reiterate the contrast with harmful disturbances caused by human activities such as conversion of forests to cities or farms, habitat fragmentation, and introduction of harmful invasive species. This class of disturbances nearly always has negative effects on butterfly diversity; Dirzo, Young, Galetti et al. 2014, "Defaunation in the Anthropocene."

Page 206: See Haddad, N. M., 2018, "Resurrection and resilience of the rarest butterflies," *PLoS Biology* 16:e2003488.

Page 215: See Haddad, N. M., L. A. Brudvig, J. Clobert et al., 2015, "Habitat fragmentation and its lasting impact on earth's ecosystems," *Science Advances* 1(2):e1500052.

ILLUSTRATION CREDITS

FIGURE 1.1A. Created by Neil McCoy

FIGURE 1.1B. Created by Neil McCoy

FIGURE 1.2. Adapted by Neil McCoy from Dirzo, R., H. S. Young, M. Galetti et al., 2014, "Defaunation in the Anthropocene," *Science* 345:401–406, with authors' permission

FIGURE 2.1. Created by Neil McCoy

FIGURE 7.1. Photo courtesy of the State Archives of Florida

FIGURE 7.2. Photo taken by Flip Schulke

FIGURE 9.1. Adapted by Neil McCoy from Agrawal, A. A., and H. Inamine, 2018, "Mechanisms behind the monarch's decline," *Science* 360:1294–1296, with authors' permission

FIGURE 10.1. Adapted by Neil McCoy from Haddad, N. M., 2018, "Resurrection and resilience of the rarest butterflies," *PLoS Biology* 16:e2003488. Photos (top to bottom): Elizabeth J. Evans/Fort Bragg, Bobby McKay, public domain

PLATE 1. Top: Photo by Nick Haddad. Bottom: Photo by Kim Davis, Mike Stangeland, and Andy Warren, provided courtesy of the Butterflies of America Foundation

PLATE 2. Photo by Stuart Weiss

PLATE 3. Top: Photo by Stuart Weiss. Bottom: Photo by Stuart Weiss

PLATE 4. Photo by Cheryl Schultz

PLATE 5. Top: Photo by Brian Hudgens. Bottom: Photo by Cheryl Schultz

PLATE 6. Photo by Randy Newman

PLATE 7. Top: Photo by Randy Newman. Bottom: Photo by Nick Haddad

PLATE 8. Photo by Johnny Wilson

PLATE 9. Top: Photo by Molly McCarter. Middle: Photo by Nick Haddad. Bottom: Photo by Martha Reiskind

PLATE 10. Photo by Jenny McCarty

PLATE 11. Top and bottom: Photos by Nick Haddad

PLATE 12. Photo by Jaret Daniels

PLATE 13. Top: Photo courtesy of the Florida Museum of Natural Sciences. Bottom: Photo by Nick Haddad

PLATE 14. Photo courtesy of the British Museum of Natural History

PLATE 15. Photo by Jeff Pippen

PLATE 16. Photo by David Pavlik

BUTTERFLY IMAGES AT START OF EACH CHAPTER:

CHAPTERS 2–7 AND 9. Images adapted from photos by Kim Davis, Mike Stangeland, and Andy Warren, provided courtesy of the Butterflies of America Foundation

CHAPTER 8. Image adapted from photo courtesy of the British Museum of Natural History

INDEX

NOTE: Page numbers followed by *f* indicate a figure.

American Beaver *(Castor canadensis)*, 120
American Chaffseed *(Schwalbea americana)*, 129
ant mutualists
　of British Large Blue caterpillars, 175–83
　of Fender's Blue caterpillars, 48, 59
　of Miami Blue caterpillars, PLATE 9
assisted migration, 229n
Atrytonopsis new species 1. *See* Crystal Skipper
Austin, George, x, 221n

Bahamian Swallowtail *(Papilio andraemon)*, 166
Ball, Brian, 120, 122, 127–28, 134, 139
Balloon Vine, 108
Baltimore Checkerspot *(Euphydryas phaeton)*, 13
Bartram's Scrub-Hairstreak *(Strymon acis bartrami)*, 61, 107, 185, 205
Bates, Henry, 222n
Bay Checkerspot *(Euphydryas editha bayensis)*, 8, 21–40, PLATE 2
　caterpillar of, 25–26, 35, PLATE 3
　climate extremes and, 29–32, 36–37
　declines of, 22–24, 203
　discovery of, 24–25, 225n
　egg laying of, 25, 29, 35
　environmental change and, 23–24, 26–27
　fragmented habitat of, 22, 32–33, 205–6
　habitat restoration for, 28, 31–35, 38–39, 208, 213
　host plants of, 25–26, 31
　nitrogen pollution and, 27–28, 34–35, 205
　restoration of, 31–40, 52, 92
　serpentine grasslands habitats of, xi, 24–27, PLATE 3
　single population of, 9, 10*f*, 32–34, 39–40
　total number of, 10*f*
Beach Vitex *(Vitex rotundifolia)*, 74
beavers, 120, 126, 131–37, 208, 231n
Behr, Herman, 15
Behren's Silverspot *(Speyeria zerene behrensii)*, 21
Bethune-Baker, George, 85–86
biodiversity. *See* diversity
Black Gum *(Nyssa sylvatica)*, 121
Black Rhinoceros *(Diceros bicornis)*, 2, 222n
the blues, 85–86, 226n
　See also British Large Blue; Fender's Blue; Miami Blue

British Large Blue *(Maculinea arion eutyphron)*, 173–86, PLATE 14
   ant mutualists of, 175–83
   collecting of, 175
   declines of, 174–75, 179–80, 182–83
   discovery of, 173–74
   extinction and reestablishment of, 11–12, 15, 173, 183–86, 207f
   habitat disturbance and, 152, 181–83, 207f, 208–9
   habitat loss and, 174, 184
   habitat restoration and, 179–83
   historic range of, 174
   host plant of, 174, 175, 179
   lab-raised populations of, 182–83, 213
   number of populations of, 10f
   rabbits and, 181–82
   total number of, 10f
Brower, Lincoln, 191, 197
Brown, Larry, 150
Burns, John, 65–66, 82–83
butterflies. *See* common butterflies; rare butterflies
butterfly collecting, 117–19, 148–50, 160–61, 175, 180, 231n
butterfly life cycles, 8

Calephelis tikal, x
Callippe Silverspot *(Speyeria callippe callippe)*, 21
Cannon, Paula, 95
captive-reared butterflies, 57–58, 161–62, 213–14
   of British Large Blue, 182–83
   mating of, 140
   of Miami Blue, 92–93, 96–97, 110
   of Schaus' Swallowtail, 161–65
   of St. Francis' Satyr, 138–42
Carolina Satyr *(Hermeuptychia sosybius)*, 114
Cassius Blue *(Leptotes cassius)*, 85–86, 95
caterpillars (larvae)
   ant mutualists of, 48, 59, 175–83, PLATE 9
   of Bay Checkerspot, 25–26, 35, PLATE 3
   of Fender's Blue, 48–49, PLATE 5
   five stages (instars) of, 8
   of Miami Blue, 86–87, 92, 100, 104–5, PLATE 9
   of St. Francis' Satyr, 123, 125, 140, PLATE 11
   temperature fluctuations and, 30–31
   transport and release of, 35–36, 52, 58
Cayton, Heather, 130
Center for Biological Diversity, 197
Center for Food Safety, 197
Ceraunus Blue *(Hemiargus ceraunus)*, 85–86
Chapman, Thomas, 176
Cheese Shrub *(Morinda royoc)*, 146
Cheney, Robert, 221n
chrysalis, 8
citizen science, 195–97, 210–11, 237n
climate change, 3, 57, 205
   drought/rainfall extremes and, 29–30, 39, 104–5, 126, 146–47, 153, 165, 205
   hurricanes and, 94, 97–99, 105–7, 157
   sea-level rise and, 81–82, 106–7
   temperature fluctuations and extremes of, 30–31, 130, 196, 205, 237n
   See also environmental change; habitat loss
Common Buckeye *(Junonia coenia)*, 119
common butterflies, 2, 112
   declining numbers of, 17–18, 203–4
   diversity of, 205
   number of populations of, 9
   See also Monarch
common subspecies, 9
Common Velvetgrass *(Holcus lanatus)*, 43
Common Wild Oat *(Avena fatua)*, 24
Comstock, William, 86
conservation biology, 22, 153

Convention on International Trade in Endangered Species (CITES), 6
Cottonmouth *(Agkistrodon piscivorus)*, 121, 122
Covell, Charles, 147, 151, 153–54, 233n
Crone, Elizabeth, 59–60
Crystal Skipper *(Atrytonopsis quinteri)*, 65–84, 185, PLATE 6
  barrier islands (NC) habitat of, 66–72, plate 7
  caterpillar of, plate 7
  classification and naming of, 65–66, 83, 228n
  conservation plans for, xi–xii, 75–78
  declines of, 72–75, 78–79
  dispersal over time of, 77–78
  habitat degradation and, 72–75, 78–79
  habitat restoration for, 73–75, 82–84, 215, 228n
  host plants of, 66, 72, 74, 79–83
  number of populations of, 10*f*
  official protection of, 66–68
  population sizes of, 77
  range of, 71–72
  reproduction rates of, 66
  restoration and recovery of, 78–83
  sea-level rise and, 81–82
  total number of, 10*f*
  tracking of, 75–77, 228n
Curry-Pochy, Keith, 166–67
Cushman, Hall, 226n

Dakota Skipper *(Hesperia dacotae)*, 193, 204
Daniels, Jaret, 199–200
  Miami Blue restoration and, 86–87, 91–94, 96, 108, 110–11
  Schaus' Swallowtail restoration and, 152–53, 159, 162, 164–65
Davis, Tad, 113–14
Denseflower Indian Paintbrush *(Castilleja densiflora)*, 26
diapause, 8

Dirzo, Rodolfo, 17
disturbance. *See* habitat disturbance
diversity, xi, xii, 15, 39, 168, 205
Dune Prickly Pear *(Opuntia pusilla)*, 70
Dusted Skipper *(Atrytonopsis hianna)*, 66–67
Dwarf Plantain *(Plantago erecta)*, 25–26

Eastern North American Monarch *(Danaus plexippus plexippus)*. *See* Monarch
Eastern Pondhawk *(Erythemis simplicicollis)*, 142
"Ecology and Conservation of Fender's Blue Butterfly" (Schultz), 53, 227n
Edwards, William Henry, 13
Ehrlich, Paul, 22, 29, 225n
endangered species, 2–5, 12–14, 129–30, 222nn
  butterfly collecting and, 117–19, 160–61, 175, 180, 231n
  habitat loss of, 3–4, 7, 16–18
  legal definition of, 6
  legal protections of, 14
  official list of, 5–6, 21–22, 61, 96, 116–19, 152–53, 223n, 233n
  See also rare butterflies; threatened species
Endangered Species Act of 1973, 5–6, 151–52
  criteria for protection by, 197–98
  official list of, 6, 21–22, 96, 116–19, 152–53, 222n
  restrictions imposed by, 96
environmental change, 4–6, 11, 13–14
  Bay Checkerspot and, 23–24, 26–27
  British Large Blue and, 174
  Crystal Skipper and, 72–75, 79, 81
  Miami Blue and, 94, 97–99
  Monarchs and, 16–17
  natural forms of, 4, 94
  Schaus' Swallowtail and, 153–55
  urban development and, 144, 147–48, 153–55, 233n
  See also climate change; habitat disturbance; habitat loss

European Rabbit *(Oryctolagus cuniculus)*, 181–82
extinct subspecies, 9, 14–18, 17–18, 200, 237–38n
  butterfly collecting and, 117–19, 160–61, 231n
  habitat losses and, 3, 7, 15–18, 42–44, 205
  hurricanes and, 88–89
  population size and, 53–54, 227n
  reestablishment of, 11–12, 15–16, 31–40, 183–86
  See also British Large Blue; Fender's Blue; Miami Blue; St. Francis' Satyr

Fender, Kenneth, 41
Fender's Blue *(Icaricia icarioides fenderi)*, 41–64, PLATE 4
  ant mutualists of, 48, 59
  apparent extinction of, 41–42, 63, 207*f*
  caterpillars of, 48–49, 58, PLATE 5
  chance of survival of, 54
  conservation-reliance of, 63
  discovery of, 41
  egg laying of, 59
  fire and, 42–44, 58–62
  flight behavior of, 55–58
  habitat disturbance and, 152, 207*f*, 208–9, 216
  habitat restoration for, 49–52, 213, 215
  host plants of, 43, 45–47, 50–52, 209
  number of populations of, 9, 10*f*, 52
  Pacific Northwest prairie habitat of, 42–44, 47, PLATE 5
  population size of, 53–54, 227n
  range of, 49, 52–53
  rediscovery of, 44–47
  restoration of, 52–64
  total number of, 10*f*, 53, 60–62
fire, 4, 51
  in controlled burns, 59, 133, 136
  habitat restoration and, 58–62, 126–27, 132–33, 208

  suppression of, 4, 13, 42–44, 132, 133
Flockhart, Tyler, 236n
Florida Keys Blackbead *(Pithecellobium keyense)*, 100–101, 106–7, 110–11, PLATE 9
Florida Strangler Fig *(Ficus aurea)*, 145
Frohawk, Frederick, 176–78

Garden (Zipper) Spider *(Argiope aurantia)*, 142
Gemmed Satyr *(Cyllopsis gemma)*, 114
gene drive, 160, 233n
Georgia Satyr *(Neonympha areolatus)*, 114
Giant Panda *(Ailuropoda melanoleuca)*, 2, 222n
Giant Swallowtail *(Papilio cresphontes)*, 166
Goss, Herbert, 174–75
greenhouse-raised butterflies. *See* captive-reared butterflies
greenhouses, 57–58
Green Iguanas *(Iguana iguana)*, 98–99, 107
Grimshawe, Florence, 148–49, 157, 160, 232n
Grinnell, Marc, 118
Guava *(Psidium guajava)*, 146
Gumbo Limbo *(Bursera simaruba)*, 145

habitat disturbance, 58–62, 120, 126–27, 130–37, 141, 152, 205–9, 212, 214–16, 231n
  from fire, 4, 58–62, 126–27, 132–33, 136, 208
  from floods, 4, 120, 126, 131–37, 208, 231n
  harmful forms of, 206, 238n
  from herbivory, 23–24, 28, 31, 181–83
  from hurricanes, 88–89, 94, 97–99, 105–7, 145–48, 149*f*, 157–58, 229n
  metapopulation dynamics and, 135, 208
habitat loss, 3–5, 205–6

beachfront development and, 72–75, 79, 81, 153–55
beaver removal and, 133–34
extinct and endangered species due to, 3, 7, 15–18, 42–44, 205, 224n
fragmented landscapes from, xi, 22, 32–33, 71–73, 77–78, 155–57, 205–6, 228n
general causes of, 4–5, 42–44, 72–75, 87–90, 174, 184, 192–95, 205, 236nn, 238n
landscape corridors for, xi–xii, 198–99, 221n, 228n
sea-level rise and, 81–82, 106–7, 111
urban development and, 15, 23, 144, 147–48, 153–55, 205, 233n
See also environmental change; invasive species
Hall, Steve, 221n
Hamm, Chris, 11
Hammond, Paul, 45, 54
Harrison, Susan, 226n
Henderson, William, 148–49
Henry, Erica, 43–44, 56, 103–10, 166–67
Himalayan Blackberry *(Rubus armeniacus)*, 43
Hoffman, Erich, 118–19, 221n
Homer-Dixon, Thomas, 200–201
host plants, 4
 of Bay Checkerspot, 25–26, 31
 of British Large Blue, 174, 175, 179
 of Crystal Skipper, 66, 72, 74, 79–83
 of Fender's Blue, 43, 45–47, 50–52, 209
 of Miami Blue, 87, 90–93, 97–98, 100, 106–7, 110–11, PLATE 9
 of Monarch, 188, 193–94, 210–11, 235–36nn
 restoration of, 50–52, 59, 79–80
 of Schaus' Swllowtail, 145
 of St. Francis' Satyr, 122, 125, 131–32, 140, 209, PLATE 11
 See also restoration of habitat
Hughes, Phillip, 100
Hurricane Andrew, 74, 88–90, 157, 229n

Hurricane Irma, 106–7, 157–58, PLATE 9
hurricanes, 73–74, 79, 94, 99–100
 climate change and, 105–6
 Miami Blue and, 88–90, 100, 106–7, 229n, PLATE 9
 Schaus' Swllowtail and, 145–48, 149*f*, 156–58, 208

inactive state (diapause), 8
Inglorius genus, x
insects, xii, 2
 declining numbers of, 17–18, 203–4, 225n
 threats to, 204–5, 210
International Union for Conservation of Nature (IUCN) Red List, 6, 198, 205
interstate highways, 26–27, 34, 198–99
invasive species, 4, 51–52, 57, 98
 Bay Checkerspot and, 24, 28, 31, 34–35
 beachfront development and, 72–75, 79
 fire and, 43, 58–59
 herbivory of, 28, 31
 hydromechanical obliteration of, 35
Italian Ryegrass *(Lolium multiflorum)*, 24, 34

Johnson, Tracy Dice, 128

Kentucky Bluegrass *(Poa pratensis)*, 74
Kincaid, Trevor, 46
Kincaid's Lupine *(Lupinus oreganus)*, 46–52, 57, 59, 209, PLATE 5
 See also Fender's Blue
Klots, Alexander, 150
Kral, Thomas, 116–19
Kuefler, Daniel, 123
Kushlan, James, 153

Labor Day Hurricane of 1935, 89, 147–48, 149*f*, 160, 232n
lab-raised butterflies. *See* captive-reared butterflies
landscape corridors, xi–xii, 198–99, 221n, 228n

Lange's Metalmark *(Apodemia mormo langei),* 21
Large Blue subspecies, 11, 22, 92, 173–74, 183–84
Leidner, Allison, xii, 68, 71–83, 228n, PLATE 7
life cycles, 8
Linnaeus, Carolus, 173
Little Bluestem *(Schizachyrium scoparium),* 66
Little Wood Satyr *(Megisto cymela),* 114
Loammi Skipper *(Atrytonopsis loammi),* 66–67
Loftus, William, 153
Longleaf Pine *(Pinus palustris),* 121
Lotis Blue *(Lycaeides idas lotis),* 21
lupines (genus *Lupinus*), 43–47

MacGregor, E. A., 223n
Macy, Ralph, 41, 45
managed relocation, 81–82, 229n
Marsden, Herbert, 174–75
Mbashe River Buff *(Deloneura immaculata),* 15
Mediocre Skipper *(Inglorius mediocris),* ix–x, PLATE 1
metamorphosis, 8
metapopulations, 32, 33*f,* 135, 226n
  disconnections within, 156
  habitat disturbance and, 135, 208
Miami Blue *(Cyclargus thomasi bethunebakeri),* 75, 85–112, 185, PLATE 8
  ant mutualists of, PLATE 9
  apparent extinction of, 88–90, 229n
  caterpillar of, 86–87, 92, 100, 104–5, 108, 110–11, PLATE 9
  declines of, 88–90, 96–99, 108–9, 111–12, 203, 213
  egg laying of, 93
  genetic diversity of, 230nn
  host plants of, 87, 90–93, 97–98, 100, 106–7, 110–11, PLATE 9
  hurricanes and, 88–89, 94, 97–99, 105–7, 229n
  identification of, 85–86, 101–2
  lab-raised populations of, 92–94, 96–97, 110
  number of populations of, 10*f,* 88–89, 95, 107–8
  population fluctuations of, 104–5
  population size of, 91–92, 95, 102–5, 111
  range of, 87, 91, 223n
  rediscoveries of, 90–92, 94–96, 99–102, 230n
  reestablishment attempts of, 92–94
  reintroduction plans for, 108–12
  sea-level rise and, 106–7, 111
  South Florida habitat of, 87–88, 91, 230n, PLATE 9
  total number of, 10*f,* 112
milkweed *(Asclepias* sp.), 188, 193–94, 210–11, 235n
Mission Blue *(Icaricia icarioides missionensis),* 21
Mitchell's Satyr *(Neonympha mitchellii mitchellii),* 11, 116, 212, 231n
Monarch *(Danaus plexippus),* 16–17, 112, 187–201, PLATE 15
  breeding grounds of, 188–89, 236n
  caterpillars of, 193–94
  citizen science programs on, 195–97, 210–11, 237n
  declines of, 16–17, 188–92, 198, 203, 235n, 237nn
  egg laying of, 190–91
  habitat loss of, 192–95, 236nn
  host plants of, 188, 193–94, 210–11, 235–36nn
  landscape corridor (superhighway) proposal for, 198–99
  life span of, 8
  migration of, 17, 188–89, 194, 196, 237n
  neonicotinoids and, 193–94, 236n
  number of populations of, 9, 10*f*
  overwintering forest habitat of, 188–89, 192, 194, 196, 235n
  petition for threatened species designation of, 197–201
  population size of, 189–92, 235n
  range and distribution of, 187–88, 192

Roundup-ready crops and, 193, 235n
temperature extremes and, 196, 237n
total number of, 7–8, 10*f*, 187, 190–92, 198
Monarch Butterfly Biosphere Reserve, 194, 235n
Monarch Joint Venture, 196–97
Monarch Larva Monitoring Project, 195
Monarch Watch, 195
Morant's Blue *(Lepidochrysops hypopolia)*, 15
mosquito control programs, 158–60, 163, 233n
Murphy, Dennis, 226n
Myrmica sabuleti, 176, 180–81
Myrtle's Silverspot *(Speyeria zerene myrtleae)*, 21
myxomatosis, 181–82

Nabokov, Vladimir, 86, 229n
natural disturbance. *See* habitat disturbance
neonicotinoids, 193–94, 236n
Nickerbean *(Caesalpinia bonduc)*, 90–94, 97–98, 100, 110–11
Nickerbean Blue *(Cyclargus ammon)*, 85–86
nitrogen pollution, 27–28, 34–35, 205
North American Butterfly Association (NABA), 6, 195, 237n
Northern Spotted Owl *(Strix occidentalis caurina)*, 2, 222n

Oberhauser, Karen, 195, 200, 235n
Ohio Lepidopterist Society, 195–96, 237n
On the Probable Extinction of [the Large Blue] (Goss and Marsden), 174–75
Orsak, Larry, 231n
Oyamel Fir *(Abies religiosa)*, 188

Pacific Poison Oak (Toxicodendron diversilobum), 43
Painted Lady *(Vanessa cardui)*, 2, 223n
Paradise Tree *(Simarouba glauca)*, 145

Passenger Pigeon *(Ectopistes migratorius)*, 200, 237–38n
Pilotte, Ross, 125
Pineland Croton *(Croton linearis)*, 61
"Place of Sorrow: The World's Rarest Butterfly" (Grimshawe), 148
Pleasants, John, 235n
Pluer, Ben, 125
pollution, 4
    from herbicides, 193–94, 235n
    from nitrogen, 27–28, 34–35, 205
    from pesticides and insecticides, 155–56, 158–60, 163
populations, 9, 10*f*, 32–34, 53–54, 227n
    average size of, 203
    gender balance in, 167–68
    genetic diversity in, 168
    importance of, 32, 39–40
    measuring size of, 102–5, 122–25, 167, 231n
    metapopulations and, 32, 33*f*, 135, 226n
    reestablishment of, 11–12, 15, 33–40
    See also reestablished populations
Poweshiek Skipperling *(Oarisma poweshiek)*, 194, 204, 213, PLATE 16
proactive science, 213–14
Pulliam, Ron, xi
pupae, 8
Purefoy, Edward Bagwell, 176–78
Purple Owl's Clover *(Castilleja exserta)*, 25–26
Purple Pitcher Plant *(Sarracenia purpurea)*, 129
Pyle, Robert, 153

Queen Alexandra's Birdwing *(Ornithoptera alexandrae)*, 231n
Quinter, Eric, 65–66, 83

rabbits, 181–82
rare butterflies, 1–18, 202–16
    aesthetic value of, 210–11
    average population sizes of, 203
    collecting of, 117–19, 148–50, 160–61, 175, 180, 231n

rare butterflies *(cont.)*
   conservation and recovery of, 13–16, 211–16
   declining numbers of, 12, 17–18, 202–4, 222n, 224n
   discoveries of new species of, 202
   habitat disturbance and restoration for, 206–9, 212–16
   habitat loss of, 3–5, 16–18, 205–6, 224n
   identification of, 5–9, 222n
   monitoring of, 6
   most studied species of, 22
   number of populations of, 9
   resilience of, 12–14
   scientific value of, 209–10
   specialized environmental requirements of, 4–5
   subspecies of, 9–12
   total number of, 2, 7–8, 10*f*
   See also endangered species; extinct subspecies
rarest butterfly, 5–9, 167–69, 202
   See also Schaus' Swallowtail
Rawson, George, 151, 233n
Redbay *(Persea borbonia)*, 121
Red-cockaded Woodpecker *(Leuconotopicus borealis)*, 129, 209
Red List of Threatened Species (IUCN), 6, 198
reestablished populations
   of Bay Checkerspot, 31–40, 52, 92
   of British Large Blue, 11–12, 15, 183–86
   failures of, 92–94
   of Fender's Blue, 52–64
   of Miami Blue, 108–12
   of Schaus' Swallowtail, 161–69
   of St. Francis' Satyr, 130–43, 168
   See also captive-reared butterflies
reintroduction, 109
restoration of habitat, 213–16
   of Bay Checkerspot, 28, 31–35, 38–39
   of British Large Blue, 179–83
   butterfly flight behavior and, 55–58
   of Crystal Skipper, 73–75, 82–84, 228n
   diversity and, 39
   of Fender's Blue, 49–52, 55, 57–62
   focus on ecosystem in, 214–15
   invasive species removal in, 24, 28, 31, 34–35, 51–52, 57, 98
   for managed relocation, 82, 229n
   of St. Francis' Satyr, 136–38, PLATE 11
   See also habitat disturbance
Ries, Leslie, 196
Rocky Mountain Locust *(Melanoplus spretus)*, 237–38n
Rossmell, Leslie, 57
Rough-leaved Loosestrife *(Lysimachia asperulifolia)*, 129, 209
Roundup-ready crops, 193–94, 235n
Ruffin, Jane, 90
Rutkowski, Frank, 150, 152, 233n
Ryan, Sean, 223n

Saarinen, Emily, 89–90
San Bruno Elfin *(Callophrys mossii bayensis)*, 21
Saunders, Sarah, 196
Schaus, William, 144, 149
Schaus' Swallowtail *(Heraclides aristodemus ponceanus)*, 2, 8, 144–69, 186, PLATE 12
   captive breeding programs for, 161–69, 213
   caterpillar of, 145, PLATE 13
   conservation debates on, 150–52
   declines of, 150–58, 161, 163–64, 203, 233n
   disappearances and resurrections of, 89, 147–50, 168–69, 232n
   discovery of, 144
   habitat loss of, 153–57, 205
   hardwood hammock habitat of, 145–46, 153–55, PLATE 13
   host plants of, 145
   hurricanes and, 145–48, 152, 156–58, 208
   life cycle of, 145, 146

INDEX 249

mosquito control programs and, 158–60, 163, 233n
number of populations of, 9, 10f
official protections of, 152–53, 159, 160–61, 233n
overcollecting of, 150, 160–61
population sizes of, 163–65, 167–68
range of, 144
as rarest butterfly, 167–69, 202
total number of, 8, 10f, 146–47, 153, 167–69
urban development and, 144, 147–48, 153–55, 233n
Schulke, Flip, 154
Schultz, Cheryl, 47–62, 191, 227n
Scotch Broom *(Cytisus scoparius)*, 43
Scuppernong *(Vitis rotundifolia)*, 121
sea-level rise, 81–82, 106–7, 111
Seaside Little Bluestem *(Shizachyrium littorale)*, 66, 72, 74, 79–83
seasonal activity, 104
Sea Torchwood *(Amyris elemifera)*, 145
sedge *Carex mitchelliana*, PLATE 11
sedges (g. *Carex*), 121–22, 134
Severns, Paul, 44–45
Sheldon, W. G., 175
Sickle-keeled Lupine *(Lupinus albicaulis)*, 46
Silvery Blue *(Glaucopsyche lygdamus)*, 55–56
Sisk, Tom, 22
Skalski, Richard, 117–18
Slender False Brome *(Brachypodium sylvaticum)*, 43
Small Cabbage Whites *(Pieris rapae)*, 2, 204, 223n
Small Camas *(Camassia quamash)*, 43
Smilax *(Smilax rotundifolia)*, 121
Smith, Charles, 46
Soft Brome *(Bromus hordeaceus)*, 24
solarization, 51
spatial ecology, 22, 225n
species (def.), 9
Spicebush Swallowtail *(Papilio troilus)*, 119

Spurred Lupine *(Lupinus arbustus)*, 45–46
Steele Cabrera, Sarah, 107–11
Sternitzky, Robert, 24, 225n
St. Francis' Satyr *(Neonympha mitchellii francisci)*, xi–xii, 11, 75, 113–43, 185, PLATE 10
apparent extinction of, 118–19, 207f
army's role in, 1, 5, 80, 113–18
artillery fire and, xii, 118–19, 127–33, 215
caterpillar of, 123, 125, 140, PLATE 11
declines of, 124–27, 203
discovery of, 114–16
egg laying of, 125–26, 139–40
grassy wetland habitat of, 1, 114–15, 119–22, 129–30, 134–35, 221n
habitat disturbance and, 126–27, 130–38, 141, 152, 207f, 208–9, 213, 231n, PLATE 11
host plant of, 122, 125, 131–32, 140, 209, PLATE 11
lab-raised populations of, 138–42
life span of, 8, 122–24
newly discovered populations of, 122–25
number of populations of, 10f, 119, 120, 122, 127, 130
official protection of, 116–18, 119, 230n
population size of, 122–26, 142–43, 230n, 231n
predators of, 142
range of, 130, 142–43
restoration of, 130–43, 168, 199
sedentary behavior of, 122, 139–40
total number of, 10f, 117
Sthenele Satyr *(Cercyonis sthenele sthenele)*, 205
subspecies, 9–12, 222n, 224nn
definition of, 9, 223n
interbreeding potential of, 9
significance of, 11–12
Swedish Large Blue *(Maculinea arion arion)*, 11, 183–84

Sweet Pitcher Plant *(Sarracenia rubra)*, 129
Systema Naturae (Linnaeus), 173–74

Tag Alder (Alnus serrulata), 121
Tall Oatgrass *(Arrhenatherum elatius)*, 43
Taylor, Chip, 195
Taylor's Checkerspot *(Euphydryas editha taylori)*, 80
Thomas, Jeremy, 180
threatened species
    butterfly collecting and, 175, 180
    legal definition of, 6
    official list of, 6, 21–22, 38, 61, 152–53, 223n
    petition for Monarch's designation as, 197–201
    See also endangered species; rare butterflies
Trans-Mexican Volcanic Belt, 235n
Tropical Storm Nicole, 99–100

United Kingdom Butterfly Monitoring Scheme (UKBMS), 6
US Fish and Wildlife Service threatened and endangered species list, 6, 21–22, 38, 61, 222n
US Forest Service, xi, 221n

Variegated Fritillary *(Euptoieta claudia)*, 119
Venus Flytrap *(Dionaea muscipula)*, 129, 209

Warchola, Norah, 59
Warren, Andy, 15
Weiss, Stuart, 27–28, 31–39
Western North American Monarch *(Danaus plexippus plexippus)*, 191–92
    See also Monarch (Danaus plexippus)
Wild Lime *(Zanthoxylum fagara)*, 145
Wild Thyme *(Thymus praecox)*, 174, 175, 179
Wilmers, Tom, 95, 100–102
Wilson, Johnny, 99–100
Wiregrass *(Aristida stricta)*, 121

Xerces Blue (Glaucopsyche xerces), 15, 205
Xerces Society, 197

Yellow Pitcher Plant *(Sarracenia flava)*, 129
Young, Frank, 150

Zipkin, Elise, 196